DEALING WITH WASTE

DEALING WITH WASTE

Resource recovery and entrepreneurship in informal sector solid waste management in African cities

Edited by

Onyanta Adama and Thaddeus Chidi Nzeadibe

AFRICA WORLD PRESS

TRENTON | LONDON | CAPE TOWN | NAIROBI | ADDIS ABABA | ASMARA | IBADAN | NEW DELHI

AFRICA WORLD PRESS
541 West Ingham Avenue | Suite B
Trenton, New Jersey 08638

Book design: Dawid Kahts
Cover design: Ashraful Haq

Library of Congress Cataloging-in-Publication Data
Names: Adama, Onyanta, editor. | Nzeadibe, Thaddeus Chidi, editor.
Title: Dealing with waste : resource recovery and entrepreneurship in informal sector solid waste management in African cities / edited by Onyanta Adama and Chidi Nzeadibe.
Description: Trenton : Africa World Press, 2016. | Includes bibliographical references and index.
Identifiers: LCCN 2016023145| ISBN 9781592215270 (hb : alk. paper) | ISBN 9781592215287 (pb : alk. paper)
Subjects: LCSH: Refuse and refuse disposal--Africa. | Municipal services--Africa. | Informal sector (Economics)--Africa.
Classification: LCC HD4485.A35 D43 2016 | DDC 363.7285096--dc23
LC record available at https://lccn.loc.gov/2016023145

Contents

Acknowledgement

The main inspiration for writing this book comes from a workshop on *Informal Solid Waste Management in Nigeria* held in Abuja on August 12, 2012. The event brought together academics as well as representatives of the government, international organizations, the formal and informal private sector, NGOs, and professional organizations to discuss topical issues related to solid waste management but with a specific focus on informal sector solid waste management. The topics discussed ranged from the relevance of the sector to poverty reduction, the ability to build social networks, relations with the state as well as age and gender dynamics. We would like to thank all those who attended and actively participated in the workshop. Our special thanks to the Swedish International Development Agency (SIDA) for funding the workshop.

At the time of writing this book, one of the editors, Dr Onyanta Adama was a researcher at the Nordic Africa Institute, Uppsala, Sweden. She would like to thank the institute for funding the book project. She is particularly grateful to the head of administration, Tania Berger for facilitating the aforementioned workshop and to the research communicator, Johan Sävström for actively participating in the workshop and facilitating the book project.

Finally our special thanks to all those who contributed chapters. The book is the product of their hard work and dedication. We would especially like to honour the contribution of Prof. Raymond N.C. Anyadike who unfortunately passed on before the publication of the book as the last sentence of the paragraph and acknowledgement.

Acronyms

AACA	Addis Ababa City Administration
AAU	Addis Ababa University
ARF	Advance Recycling Fee
BAN	Basel Action Network
BATRRT	Best Available Treatment Recovery and Recycling Techniques
BFR	Brominated Flame Retarded
BPR	Business Process Reengineering
CASSAD	Centre for African Settlement Studies and Development
CCC	communal container collections
CH_4	Methane
COPD	Chronic obstructive pulmonary disease
CV	Calorific value
CWG	Collaborative Working Group on Solid Waste Management in Low-and Middle-income Countries
DEFRA	Department for Environment, Food and Rural Affairs
EEESL	Emoc Engineering and Environmental Services Limited
ENDA	Environmental Development Action in the Third World
EoL	End-of-Life
EPA	Ethiopia, Environmental Protection Authority
EPR	Extended Producer Responsibility

EPZ	Ghana Free Zone
FDA	French Development Agency
FDA	French Development Agency
FEPA	Federal Environmental Protection Agency
GcNET/GCMS	Ghana customs, excise and prevention services data management system
GIZ	Deutsche Gesellschaft für Internationale Zusammenarbeit
GTZ	GesellschaftfürTechnischeZusammenarbeit
HH	house-to-house
IGNIS	Sustainable Valorisation of Municipal Solid Wastes in Emerging Mega Cities
ILO	International Labour Organization
ISSWM	Informal sector solid waste management
ISW	Informal solid waste
ISWA	International Solid Waste Association
ISWM	Integrated solid waste management
ISWMS	Integrated Solid Waste Management Strategy
IWM	Informal Waste Management
KIIs	Key-Informant Interviews
LAWMA	Lagos State Waste Management Authority
LGAs	Local Government Areas
LSRDB	Lagos State Refuse Disposal Board
LSWDB	Lagos State Waste Disposal Board
LWCPs	Local waste collection points
MDGs	Millennium Development Goals
MEEP	Ministry of the Environment and Physical Planning
MoE	Ministry of Environment
MSW	Municipal Solid Waste
MSWM	Municipal Solid Waste Management
MTTU	Ghana Police Motor Transport and Traffic Unit
NFPO	National Forum of Peoples Organisations
NGOs	Nongovernmental organizations
OAU	Organization of African Unity
ODA	Official Development Assistance
OECD	Organisation for Economic Co-operation and Development
OSIWRP	Ondo State Integrated Waste Recycling Plant

PCBs	Polychlorinated Biphenyls
PDF	Pre-Disposal Fee
POPs	Persistent Organic Pollutants
PPIAF	Public Private Infrastructure Advisory Facility
PPP	Public Private Partnership
PRO	Producer responsibility organisations
PSP	Private sector participation
SAP	Structural Adjustment Programme
SENSE	Swiss Foundation for Waste Management
SWICO	Swiss Association for the Information, Communication and Organisational Technologies
SWMAs	Solid waste management agencies
UEEE	Used Electrical and Electronic Equipment
UNCHS	United Nations Human Settlements Programme
UNEP	United Nations Environment Programme
UNICEF	United Nations Children's Fund
WEEE	Waste Electrical and Electronic Equipment
WEP	Women Environment Programme
WIEGO	Women in Informal Employment: Globalizing and Organizing
WSSD	World Summit in Sustainable Development

Foreword

This book is a collection of studies which bring together in one volume very current and exciting views on problems of urban solid waste management in developing world cities with special reference to the informal sector in Africa. Over time, cities worldwide have contended continuously with the search for an appropriate response to the ever-escalating problems posed by urban solid waste management. In reality, it is a complex phenomenon with differential response depending on the extent to which the expedient legislative, technological and pecuniary resources, and requisite planning and innovation are able to respond to increasing volumes over time.

Indeed, escalation in collection and recycling inadequacies with increasing urbanization is the norm. In cities of developed countries, public-private sector collaboration, government policy and legislation, especially proactive legally binding targets for recycling, composting and landfilling are deployed to contend with the problem. This is in addition to massive private sector investment in technology and equipment, continuous innovation as in the area of waste to fuel and waste to energy programmes. These have all combined to ensure the affirmation of the standing on formality.

Not so in cities of the developing world. Bereft of the underpinnings which promote and sustain formality and in the face of unavailability of centralised waste collection and treatment and other innovations, the informal system has been making steady gain in the last two decades. The pull factors include the increasing availability of informal sector candidates thrown up by the rising tide of the phenomenon of rapid urbanization of poverty typical of African cities, ease of access into the informal universe , increasing unserviced areas and rising volume of diversified recyclables in the urban waste stream. The outcome of increasing engagement of waste pickers has been its salutary impact on urban livelihood and poverty reduction.

In this remarkable publication, the authors provide us with an in-depth and critical discourse on the informal sector which captures many angles of the phenomenon in Africa. The Nigerian space with the highest single concentration of highly diversified cities has been retained as an opposite pool from which the authors selected representative examples to expound the wide-ranging aspects of the informal waste sector. Two of the chapters delved into the informal sector from a national and subnational scales: informal recycling market linkages and urban governance in Nigeria; and Almajiri and waste management in urban centres in Northern Nigeria. At the city level, the authors examined dynamics of informal solid waste recycling in Zaria and children's participation in informal solid waste management in Kaduna.

Two studies focused on Abuja, the Federal Capital Territory (FCT) and Lagos namely: informal waste recycling and private sector in integrated solid waste management. Outside Nigeria, the city of Addis Ababa, capital of Ethiopia received attention with a perspective on the contributions of international organizations in solid waste management while chapter another interrogates how the opportunities gleaned from the relationship between the informal waste collectors and some steel industries can help transform traditional waste management practice toward zero waste practice in Accra, Ghana. The rest of the chapters examined material and energy recovery potentials of municipal waste in Nigeria; scoping extended producer responsibility for e-waste problem; and waste recycling and poverty alleviation in Nigeria. Against this backdrop, the preponderance of contributions from Nigerian scholars, mostly home-based with a complement of a few from abroad is to be expected.

The book is equally remarkable because this is perhaps the first time such attention is devoted to informal solid waste management in African countries in a single volume. The authors are also accomplished and respected scholars and professionals with sterling contributions in diverse areas of urban solid waste management and especially in informal sector.

The prestigious Nordic Africa Institute (NAI) Uppsala, the international centre of research, documentation and networking provided the requisite academic platform for the two editors: Uppsala-based Onyanta Adama, Senior Researcher at NAI and Chidi Nzeadibe , Visiting Scholar and Senior Lecturer from the University of Nigeria, Nsukka to accomplish their task with aplomb. Like most of the other contributors, I have been captivated overtime by Dr. Adama's insightful research publications on diverse aspects of informal solid waste management founded on her primary research interests of urban governance and service delivery in African cities. My connection with Dr. Nzeadibe which dates back to his undergraduate years in the Department of Geography, University of Nigeria, Nsukka, Nigeria in the 1990s was strengthened by my facilitating his acceptance at the then Federal Environmental Protection Agency Zonal Office in Port Harcourt. There he developed an abiding interest and passion in urban solid waste management which he followed up with specialization in solid waste management and recycling in his graduate programme. As a staff since 2006, Dr. Nzeadibe has researched and published extensively in reputable international journals and participate actively in research affiliations and communities of practice revolving around solid waste management.

The book amply illustrates the importance of this field by reason of the wealth of information, sources and observations it provides. Through their pages, one will discover some of the most exciting frontiers in informal solid waste management as well as opportunities. I think that the authors can be confident that there will be grateful readers who would have gained a broader perspective in the field as a result of their efforts. Their book is destined to play a major role in exciting , motivating and educating a broad audience of present and future generations of students, academics , researchers , NGOs, policy makers and international development agencies worldwide especially in the US, Europe and Africa. Although the emphasis of the book is informal solid waste management, it contains much that will be of interest to those outside the field and indeed to anyone with interest in urban economy and environmental quality, poverty alleviation and urban livelihood.

Prof. James C. Nwafor,
Shell-Centre for Environmental Management and Control (CE-MAC), University of Nigeria, Enugu Campus
&
Executive Director, Environment and Development Policy Centre for Africa (EDPCA), Enugu,
Enugu State, Nigeria.
Email: jamescnwafor@yahoo.com 31st October 2014.

INTRODUCTION: RESOURCE RECOVERY AND ENTREPRENEURSHIP IN INFORMAL SECTOR SOLID WASTE MANAGEMENT IN AFRICAN CITIES

Onyanta Adama and Thaddeus Chidi Nzeadibe

Informal Sector Solid Waste Management (ISSWM) dates back many centuries but its rising profile in Africa is commonly attributed to the urban crisis that followed the Structural Adjustment Programmes of the 1980s. With many African economies in stagnation, the poor were largely left to devise their own strategies for survival.

ISSWM consists of two major components (Gunsilius et al. 2011). The first is the provision of informal services, largely waste collection, in areas left out of the formal system. The second refer to the individuals, cooperatives, family and micro-enterprises that function mainly as an extractive industry. Recycling, which involves collecting, sorting, processing, storing and trading waste materials into the industrial value chain, falls into the second category. While ISSWM is common in African cities, there is a dearth of literature on the subject. Much of the available research relates to Asia and

Latin America. Except for the celebrated case of the *Zabaleens* in Cairo, Egypt and more recently, the reclaimers of South Africa, Africa has received very little attention in international debates on informal sector involvement in Municipal Solid Waste Management (MSWM). In August 2012, the Nordic Africa Institute organised a workshop on informal solid waste management in Nigeria, which attracted academics from several universities as well as representatives of the formal and informal private sector and NGOs. The presentations covered major topical issues, including environmental sustainability, age and gender dynamics, social networks and the formal/informal interface. This publication is an outcome of the workshop.

While the poverty reduction potential of ISSWM is acknowledged, informal activities do not occur in a vacuum. There also is a need to draw attention to the broader environmental, political, economic and sociocultural issues that shape the sector. Highlighting the dynamism and complexity of ISSWM in African cities is a primary aim of this book.

Context

The concept of "informal sector" or, in more recent times, "informal economy," remains a major topic of interest in African urban and development studies. The interest mainly stems from the persistence of the phenomenon and a longstanding debate on the definition, applicability and relevance of the concept. The International Labour Organization (ILO) defines the informal sector as comprising activities characterised by ease of entry; reliance on indigenous resources; family ownership; small-scale operation; labour intensiveness and adapted technology; skills acquisition outside the formal school system; and unregulated and competitive markets (Bangasser 2000). In 1993, the International Conference of Labour Statisticians defined the informal sector as a group of household enterprises or household-owned unincorporated enterprises that includes informal own-account enterprises and enterprises of informal employees (International Labour Organization 2002). The ILO definition has been criticised for making assumptions and neglecting the complexity and heterogeneity of informal sector activities (Myers 2011; Trager 1987). For example, it is now commonly acknowledged that the informal economy is highly differentiated and traversed by hierarchies, divisions and inequalities structured along lines of gender, age, ethnicity, race and income, among others (Lindell 2010). Other issues requiring attention are the formal/informal interface and linkages with the formal private sector and other actors beyond the urban scale.

Historically, the informal has been defined largely in opposition to the formal. In practice, the boundaries are not often clear. For example, there is a common perception that those working in the informal economy avoid regulation and therefore operate illegally. However, not only is the line between legality, illegality and criminality blurred, there are questions as to who or what remains unregulated and why (Alter Chen 2005). To move the debate on informality forward, recent writing on informality in Africa advocates going beyond the formal/informal dichotomy to a more nuanced understanding (Myers 2011). A major argument is that the informal is not outside the formal, and that processes of formalisation and informalisation work together (Mbembe and Nuttall 2004 cited in Myers 2011). In summing up, as Yusuf (2011) points out, existing theoretical approaches do not adequately capture the internal diversity, dynamism and regulatory processes of the informal economy in developing countries, and especially in Africa.

The informal waste sector is defined as comprising individuals or small and micro enterprises that become involved in waste management without being registered or formally charged with providing services (Medina 2011). In line with the debate on informality above, it is important to acknowledge that ISSWM is complex. For example, the formal/informal divide is not so clear cut. It has been observed that cooperatives formed by informal workers undertake work under contract from municipal authorities (Gunsilius et al. 2011). In terms of legality, and contrary to the common perception, many informal waste workers and businesses are registered with state authorities and pay taxes (Velis et al. 2012). While these issues are now being debated, the poverty reduction potential of ISSWM has received most of the attention. Due to the relative ease of entry, and access to a continuous supply of solid waste resulting from continued poor management by the authorities, ISSWM is considered one of the most popular livelihood activities. Thousands of urban dwellers across the globe make a living from recovering, sorting and selling recyclable materials. Almost all the chapters in this book acknowledge the poverty reduction potential of ISSWM, but there is a need for greater understanding of broader issues such as the environmental implications, the linkages and relations with other sectors and actors across different spatial scales, the market dynamics and the social relations of power as the various groups try to maximise their opportunities.

Environmental Sustainability

The Stockholm conference on Persistent Organic Pollutants (POPs) in May 2001 was the first major global attempt to draw attention to the link between improper solid waste management and climate change. The latter is largely attributed to greenhouse gas emissions (McGranahan 2007) and the waste sector is a noted contributor to global anthropogenic emissions, a major cause of climate change (Monni et al. 2006). Currently, the world generates about 1.3 billion tons of waste per year, a figure that is set to increase to 2.2 billion by 2025 (Hoornweg and Bhada-Tata 2012). Increasing waste generation means an equal increase in the rate of gas emissions and other environmental stressors, such as land reclamations for landfills. Emissions from landfills, including methane and carbon dioxide will increase, in addition to emissions from incineration and open burning of waste (Monni et al. 2006). The toxins released affect ecosystems, contaminate groundwater sources and add to pollution. In many cities in the developing world, waste minimisation and reduction, source separation and composting are not common (Adama 2007). Open dumping, which leads to pollution of aquifers and water bodies, and burning of waste, which is accompanied by air pollution, are widespread and recycling is largely informal (Hoornweg and Bhada-Tata 2012). Unlike in developed countries, where the main motivation for recycling is environmental protection, in developing countries, recycling is largely a private economic activity. However, there are positive implications for the environment, albeit by default since the main motivation for recycling is economic (Scheinberg 2011). Most of the chapters touch on the crucial issue of environmental sustainability, but the first two deals with it more explicitly. Chapter one sets out to explore the material and energy potential of municipal solid waste in Nigeria and confirms the notable contribution of the informal sector to resource recovery. It focuses primarily on the role of waste pickers. Material recovery is noted as a precursor to recycling and waste pickers play a crucial role by removing huge quantities of recyclables from the waste stream. The implications for the environment are obvious. Reducing the volume that goes for final disposal implies the lifespan of landfills is prolonged. It also suggests lower costs for government, since fewer collection vehicles and waste handling facilities are required. Chapter two continues with the theme of resource recovery and recycling, but introduces an emerging issue, the management of e-waste. The authors point to a shift by informal sector recyclers from traditional household waste to the more valuable but often hazardous electronic waste, largely driven by illegal imports from Western countries. The negative environmental and health impacts of e-waste are highlighted. In view of the limited local capacity for handling e-waste, there is a call for greater involvement by manufacturers of electronic products in e-waste management.

Market Dynamics

Informal recycling is an economic activity based on valorisation of waste (Scheinberg 2011), and entrepreneurship is a crucial factor (Rogerson 2001). Recycling involves collecting, sorting, processing, storing and trading materials into the industrial value chain. The value of a waste material increases as it moves up the chain of activities. Income levels, the existence of local and national markets, the demand for secondary raw materials, prices of materials and the level of government regulation are some of the factors that determine the extent to which a particular material is recycled (Wilson et al. 2006). Through a case study of Ibadan, Nigeria, chapter three provides useful insight into how recyclables are handled or treated and the networks that have emerged around different materials, including metal, glass, paper and wood. There is a detailed analysis of recycling practices and options, as well as an overview of the market. The authors report that recyclables are mainly sourced from mechanical workshops, factories, business and office premises and dumps. Thereafter, the materials are bought by users directly or through their agents, relatives or employees. The materials are then used by recycling companies to produce various goods for the local market or taken to other cities and sold to recycling plants. The authors also raise issues related to the importation of foreign technology and community participation.

Informal recycling is largely demand-driven and has strong links to the industrial sector (Gunsilius et al. 2011). A longstanding issue within the discourse on informality, and informal recycling in particular, is the linkages with the formal private sector. For example, the structuralist school sees the informal sector as subordinate economic units in which workers serve to reduce input and labour costs and therefore increase the competitiveness of large firms (Alter Chen 2005). The nature of capitalist development is said to account for the persistence of informal production relationships, with the formal sector keeping some facets of the informal going in order to extract profits (Portes et al. 1989 cited in Myers 2011). Chapter four interrogates the link between the informal waste sector and the steel industry in Accra, Ghana. The findings show that the informal sector provides the steel industry with cheap raw materials, which points to the subordinate nature of the informal sector. The study cites the example of a firm, Special Steel Company Limited, which uses mixed scraps to produce iron rods for local markets. Accredited agents are used by the company to establish and service linkages with the broader urban economy.

Still on linkages with the formal sector and the urban economy, but also with an interest in the linkages beyond the city scale, chapter five examines the activities of waste pickers and scrap dealers in Aba, Nigeria. The authors report that more than half the recyclables recovered by the informal sector in the city are consumed by formal sector companies. The buying and selling of recyclables occurs within highly organised systems. There are major markets for recyclables in Aba and in other Nigerian cities, but the study also notes a transborder trade. Transborder shipments of e-waste and the negative consequences for the environment were noted earlier. However, as discussed in chapter five there is a network of actors engaged in trading other recyclables across cities in different parts of the globe. These include copper, aluminium, iron, plastics and glass. The authors note the internationalisation of recycling networks between Aba and cities in neighbouring countries, including Malabo (Equatorial Guinea) and Kumba and Douala in Cameroon.

Social Dimension

Informal recycling involves a multiplicity of actors, from waste pickers to middlemen or small buyers to agents or big buyers and manufacturers. These actors are individually or collectively involved in a web of relationships in a bid to maximise their opportunities. A major outcome is a highly hierarchical sector, with waste pickers at the bottom and manufacturers at the top (Rogerson 2001). The presence of middlemen operating between waste pickers and manufacturers translates into opportunities for exploitation (Medina 2005). Waste pickers are often in highly dependent relationships and have little influence on prices of materials and network relations (Adama 2012). Sociocultural and economic factors play a crucial role in shaping the activities of actors and the social relations of power. Chapter six provides an overview of informal solid waste activities in Zaria, Nigeria. The study reports the existence of a wide range of activities, including waste collection, waste picking on streets and dumps, sorting and storage and compacting by small and large junkshops, reuse and recycling. Crucially, it also highlights the social dynamics in play. Waste picking is noted as an important survival strategy. However, the income earned is not enough to sustain the livelihoods of waste pickers. The author reports that waste pickers are often exploited by middlemen or intermediate dealers. As a result of the low income, the majority of waste pickers engage in other survival activities, including farming and petty trading. Women and children are identified as the most vulnerable.

Ethnicity, race, gender, age, religion and income influence access to the various components of the chain, as well as the ensuing power relations. There are marked gender differences in the division of labour, but with variations across cities and regions. In some parts of the world and particularly in Asia, women and girls form the majority of waste pickers (Hunt 1996), while in Nigeria waste picking is primarily a male affair (Nzeadibe 2009). Sociocultural factors largely account for these differences. Women have limited access to lucrative materials, earn less and are sometimes excluded from links in the chain on the grounds that some activities are purportedly unsuitable for women (Adama 2012). No chapter deals explicitly with gender issues, but most highlight gender differences, especially in the composition of waste pickers. Many chapters confirm that waste pickers are predominantly male.

The role of children in the informal economy has attracted much attention, in large part due to the controversy and concerns surrounding child labour. According to the ILO, about 211 million children aged between 5 and 14 are engaged in various forms of mainly hazardous work across the globe (International Labour Organization 2002). Popular activities include waste picking, street trading and home-based work. Poverty is noted as one of the most important reasons children work in the informal economy (Thorsen 2012). This accounts for the rise in the number of children involved in informal activities in times of economic crisis (International Labour Organization 2002). Furthermore, children are a source of cheap labour for large firms through outsourcing of production to developing countries. They also provide cheap labour to men and women (Thorsen 2012). In places where there are sociocultural restrictions on women's mobility, as in Kaduna, Nigeria, children are particularly useful to women because they are free to bridge the barrier that separates men and women (Coles 1991, Callaway 1987). Thus children become an important source of labour in both domestic and productive activities. For example, they care for younger siblings, run errands and hawk goods on the streets. Chapter seven provides insights into children's participation in ISSWM in Kaduna. It documents the activities children engage in and the complex forces they contend with in negotiating their survival. There is discussion of why children engage in waste picking, the time spent, and the personal benefits to them as well as to society at large. The authors also draw attention to how children negotiate urban space to gain access to recyclables, and of the market dynamics and power relations as seen through price negotiations and bargaining. Inevitably, they raise the negative consequences associated with waste picking, such as the health risks, and also discuss the issues of child labour and rights.

Chapter eight continues the theme of children in ISSWM, but focuses on a particular group, the Almajiris. These are young male pupils left by their parents with an Islamic scholar or *Mallam* to study the Koran. Several attempts have been made by the state to integrate Koranic schools into the Western educational system, but with little success. The Almajiris are arguably the most disadvantaged urban group in northern Nigeria's cities. With no parental support, ignored by government and exploited by *Mallam*s, many resort to begging, petty jobs or crime to survive. Some also engage in waste picking. In a bid to explore the potential and relevance of Almajiris in overall solid waste management, the author examines the evolution of value chains in the emerging local recycling markets and the role of recycling in job creation, waste management, peace and stability in northern Nigerian cities. The reference to peace and stability is important, given ongoing sociopolitical events in Nigeria. In recent times, Almajiris have been associated with Boko Haram, an Islamic sect that has carried out a spate of bombings in Nigeria's northern cities. They are seen as the foot soldiers of Boko Haram (Nigeria Newsworld 2012). Indeed, the challenges faced by them make them vulnerable to inducements by terror groups, but as this chapter shows, some of the more resourceful Almajiris have taken to waste picking. While Almajiris engage in waste picking primarily for survival, they are also contributing to environmental sustainability through resource recovery. The author calls for the recognition of Almajiris as valuable partners in solid waste management, a call that is not particularly challenging, since they are already involved in it. The proposed strategy requires little or no capital investment, but has great economic, social and environmental benefits for all stakeholders.

Relations with the State

ISSWM provides a unique opportunity to examine the contentious, paradoxical and constantly evolving relations between local governments and informal workers in African cities. Many of the issues centre on the capacity of the state to regulate, how regulation is carried out, the effects attempts at regulation have on relations with informal workers and the responses of these workers. ISSWM is a useful gap-filler, especially in low income areas that receive few, if any, services from the state. However, while ISSWM is a major component of urban solid waste management systems in Africa, and makes significant contributions to local economies and the environment, it is in many cases not officially recognised by governments (Nzeadibe 2009). A major problem stems from the fact that waste is owned by local governments, making informal waste activities illegal. Thus harassment of waste pickers by government officials is a common occurrence in African cities. The formalisation of informal activities is one avenue for the state to integrate the informal sector into the formal. Those who argue in favour of formalisation believe the informal sector is here to stay and requires the necessary laws, regulations and policies (Alter Chen 2005). For formalisation to take place in ISSWM, there have to be a number of enabling conditions, namely voice, visibility, validity and viability (Gerdes and Gunsilius 2010). Voice is an important first step. Informal waste workers can gain a voice by collectively organising, since this enables them to negotiate with municipal authorities. However, in many cases the majority of waste pickers remain poorly organised (Rogerson 2001).

Studies conducted in Brazil, Egypt and Costa Rica show that the informal sector can be an efficient partner of municipalities (Gunsilius 2010), but the degree of state recognition and of integration into formal solid waste management systems varies. For example, there is relatively high integration in Quezon City (Philippines), Lima (Peru) and Pune (India), while in Cairo the state is reluctant to engage with the informal sector (Scheinberg et al. 2010). Worst still, in Cluj (Romania) and Lusaka (Zambia), the state pays little attention to informal workers and is largely unaware of their activities. The last four chapters engage with the debate on the role of the state and raise specific questions. Chapter nine takes up the important issue of government recognition of informal waste workers, particularly waste pickers. As noted earlier, they are the most vulnerable group in the recycling chain. The author observes a lack of official support for waste pickers in Nigeria, even though they are the main drivers of recycling and contribute to poverty reduction. As a result of government non-recognition, efforts made so far to increase recycling rates have not yielded meaningful dividends. The author surmises that the way forward is to recognise the efforts of waste pickers and mainstream waste picking into formal systems. Chapter ten continues with the themes of recognition and formalisation. The author argues that existing activities and structures should be formalised and scaled-up to foster entrepreneurship that will support economic growth and integration of the informal sector into national, regional and global value chains. Other issues related to the role of the state noted in the discussion include the lack of comprehensive data for planning and management, the health of waste pickers and the need for appropriate policies and legislation that can drive the informal sector towards sustainable development.

Chapter eleven begins by highlighting the challenges facing formal solid waste management systems through a case study of Lagos, Nigeria. Rapid population growth and urbanisation are particularly noted. The author acknowledges the important role of the informal sector in waste collection, transportation, recovery and recycling through a presentation of the activities of cart pushers, cart builders, waste pickers, resource merchants and recyclers. In a broader context, he also lists how the informal waste sector is contributing to the achievement of the Millennium Development Goals. However, of particular interest here is the insight into how solid waste management is structured in the city. Crucially, this provides an opportunity to introduce the problems surrounding the decentralisation of services in African cities. As is the norm in cities elsewhere, local governments in Nigeria are responsible for solid waste management. However, in practice, national, state or regional governments continue to retain control (Onibokun 1999). The capital city of Abuja presents its own unique features, with the municipal council constitutionally and legally barred from handling solid waste management (Adama 2007). In Lagos, responsibility has shifted between local governments and the state ministry of environment over the years, as observed in chapter eleven. The author notes that this instability has the negative effect of denying the relevant agencies the benefit of accumulated management experience.

Largely championed by the World Bank, the term "governance" became a mantra in development discourse on Africa in the late 1980s. Urban governance is primarily concerned with the institutions and processes of city governance and with the power relations between actors (Devas 2002). The concept has been criticised for being highly normative, ahistorical, apolitical and Eurocentric, but remains the major tool for assessing and analysing infrastructure and service delivery in African cities (Myers 2011). Neoliberal approaches to governance favour minimal state involvement in service delivery and argue for greater private sector participation. The involvement of non-state actors through partnerships is a major component of governance. A notable argument in favour of partnerships is that they promote public participation and are therefore a more sustainable means of delivering urban services (Adama 2007). Public Private Partnerships, which entail the transfer of certain functions and services from state to private sector, are arguably the most common, but there are also other types involving NGOs, community groups and associations, households, donors and international organisations. Partnerships do indeed offer a way to bring state and non-state actors together to address specific problems and create new arenas for participation, but the implication is that the state has to share the institutional playing field with other actors (Allen 2004). It is therefore difficult to ignore power relations and the processes of exclusion and inclusion. As Elander (2002) points out, it is important to examine the range and balance of power in any partnership, since the actors have different capacities to influence decisions. For example, when necessary the state will rely on its authority, which confers power on it to take decisions over others (Adama 2007). This does not necessarily mean other actors are powerless. Individuals, community and civil society groups can resort to a range of strategies, from mass political action to subversion and subtle forms of resistance (Sharp et al. 2000).

Chapter twelve explores how the involvement of four international organisations has shaped solid waste management in Addis Ababa, Ethiopia. The organisations are the International Labour Organization, Income Generation and Climate Protection through the Sustainable Valorisation of Municipal Solid Wastes in Emerging Mega cities (IGNIS), the French Development Agency (FDA) and Public Private Infrastructure Advisory Facility (PPIAF). The organisations are involved in different ways, but a common aim is improved solid waste management in the city. Primarily, the author is interested in the types of networks and power relations that have emerged as a result of the involvement of these foreign organisations, and the consequences. Her study reveals that government has used its power to restrict the activities of the international organisations and was unwilling to loosen its control over civil society organisations. New laws were introduced that made it mandatory for international organisations to enter into partnerships with government agencies, but excluded other actors. Among those excluded were informal waste workers. Local NGOs are described as being mainly interested in writing project proposals for funding from the international organisations and as paying little attention to the needs and interests of the poor. International organisations are accused of failing to understand the local culture and paying little attention to the realities on the ground, including the political context.

In summing up, one must reiterate that this book deals with a sector that has important societal relevance in general, and in Africa in particular, due to its association with poverty reduction and environmental sustainability, two critical areas. As noted in the book, while the activities of informal waste workers can have negative impacts on the environment, there are also their positive contributions to the environment through resource recovery. Topics such as Extended Producer Responsibility as a tool for managing the e-waste problem and energy and material recovery from municipal solid waste are contemporary environment and development issues and have implications for informal sector involvement in MSWM. The book also confirms ISSWM as an important livelihood strategy, especially for the most vulnerable urban groups.

Within academic discourse, the book is in line with an emerging school of thought in critical urban studies aiming to counter the image of African cities as poverty-ridden, chaotic and dysfunctional. Rem Koolhaas and the Harvard School of Design's Project on the City have been notable in countering Western notions of African cities (see Gandy 2005). In trying to understand how African cities actually function and using Lagos as an example, Koolhaas draws attention to the city's morphology, the innovative skills of informal workers, the complexity of the socioeconomic structures and the series of self-regulatory systems that enable it to function, in spite of numerous challenges. The following quotes are the impressions of Koolhaas and the Harvard team on three visits to Lagos.

> Our initial engagement with the city was from a mobile position. Partly out of fear, we stayed in the car ... Lagos seemed to be a city of burning edges. Hills, entire roads were paralleled with burning embankments. At first sight, the city had an aura of apocalyptic violence; entire sections of it seemed to be smouldering, as if it were one gigantic rubbish dump...The activity taking place was not a process of dumping, but more a process of sorting, dismantling, reassembling and potentially recycling. Underneath the viaducts there was a continuous effort to transform discarded garbage (Quoted in Gandy 2005:40).

> From the air, the apparently burning garbage heap turned out to be, in fact, a village, an urban phenomenon with a highly organized community living on its crust ...(Quoted in Gandy 2005:41).

For our purposes here, the above impressions of the Harvard team are particularly relevant and interesting for their references to informal recycling. This book confirms that informal recycling is not as chaotic or dysfunctional as it seems. The activities are interconnected, which means a high level of interdependence between actors. Furthermore, informal rules and norms guide their behaviour. Underneath is a constant, complex but well-structured flow of people and materials across spatial scales (see also Adama 2010). Koolhaas and the Harvard school have been criticised for being myopic (Godlewski 2010) and essentialist (Fourchard 2011; Gandy 2006 and 2005). However, these reactions suggest that Koolhaas has succeeded in moving the debate on African cities forward. This book is a contribution to the debate. It offers an opportunity to understand how African cities function in spite of all the odds and provides unique perspectives and features. Ultimately, it also raises questions about urban governance in general and solid waste management in particular. Notwithstanding the norms of good governance that favour the formal sector, the informal continues to be a major part of urban governance systems. This calls for rethinking, or new ways of thinking, about governance in African cities.

References

Adama, O. (2007). *Governing from above: Solid Waste Management in Nigeria's New Capital City of Abuja.* Stockholm Studies in Human Geography 17. Stockholm: Almqvist and Wiksell.

Adama, O. (2010). Beyond Disfunctionality: Recycling in Kaduna, in *Annual Report 2010.* Uppsala: Nordic Africa Institute. Available at http://nai.diva-portal.org/smash/record.jsf?pid=diva2:413279

Adama, O. (2012). Urban Livelihoods and Social Networks: emerging relations in informal sector recycling in Kaduna, Nigeria. *Urban Forum* **23**(4): 449 – 466.

Allen, J. (2004). The Whereabouts of Power: Politics, Government and Space. *GeografiskaAnnaler B* 86 B(1).

Alter Chen, M. (2005). *Rethinking the informal economy: linkages with the formal economy and the formal regulatory environment.*United Nations University-WIDER Research PaperNo. 2005/10.Available at http://hdl.handle.net/10419/63329.Accessed 25 January 2013.

Bangasser, P.E. (2000). The ILO and the informal sector: An institutional history. Geneva: ILO.

Callaway, B. J. (1987). Muslim Hausa Women in Nigeria, Tradition and Change. Syracuse: Syracuse University Press.

Coles, C. (1991). Hausa women's work in a declining economy: Kaduna, Nigeria, 1980-1985. In *Hausa Women in the Twentieth Century*, C. Coles and B. Mack (eds). Madison: University of Wisconsin Press.

Elander, I. (2002).*Partnerships and urban governance.* Oxford: Blackwell.

Fourchard, L. (2011). Lagos, Koolhaas and Partisan Politics in Nigeria. *International Journal of Urban and Regional Research*

35(1): 40 – 56.

Gandy, M. (2005).Learning from Lagos. *New Left Review* **33**: 37 – 52 .

Gandy, M. (2006).Planning, anti-planning and the infrastructure crisis facing metropolitan Lagos. *Urban Studies* **43**(2): 371 – 96.

Gerdes, P. and E. Gunsilius (2010).*The Waste Experts: Enabling Conditions for Informal Sector Integration in Solid Waste Management: Lessons learned from Brazil, Egypt and India.* Eschborn: Deutsche GesellschaftfürTechnischeZusammenarbeit (GTZ).

Godlewski, J. (2010). Alien and Distant: Rem Koolhaas on Film in Lagos, Nigeria. *Traditional Dwellings and Settlement Review* **XX1**(11): 7 – 20.

Gunsilius, E., B. Chaturvedi and A. Scheinberg (2011).*The Economics of the Informal Sector in Solid Waste Management. CWG-Collaborative Working Group on Solid Waste Management in Low-and Middle-income Countries, GIZ-Deutsche Gesellschaft für Internationale Zusammenarbeit (GIZ) GmbH.*

Gunsilius, E. (2010). Role of Informal Sector in Solid Waste Management and Enabling Conditions for its Integration.Experiences from GTZ. Paper presented at Transwaste workshop on the informal sector, April 4. Available at http://www.transwaste.eu/file/001441.pdf.Accessed 18 May 2012.

Hoornweg, D and P. Bhada-Tata (2012).*What a Waste: A Global Review of Solid Waste Management.* Washington DC: World Bank. Available at http://www-wds.worldbank.org/external/default/WDSContentServer/WDSP/IB/2012.Accessed 30 January 2013.

Hunt, C. (1996). Child waste pickers in India: The occupation and its health risks. *Environment and Urbanization* **8**: 111 – 18.

International Labour Organization (2002).*Decent work and the informal economy*. Geneva: ILO Publications.

Lindell, I. (2010). Introduction: The changing politics of informality – collective organizing, alliances and scales of engagement.

In *Africa's Informal Workers: collective agency, alliances and transnational organizing in urban Africa,* I. Lindell (ed.). London/Uppsala: Zed Books and Nordic Africa Institute.

Mbembe, A. and S. Nuttall (2004).Writing the world from an African metropolis. *Public Culture* **16**(3): 347 – 372.

McGranahan, G. (2007). *Urban environments, wealth and health: Shifting burdens and possible responses in low and middle-income nations.* Human Settlements Discussion Paper Series. London: International Institute for Environment and Development.

Medina, M. (2005).Waste Picker Cooperatives in Developing Countries. Paper prepared for WIEGO/Cornell/SEWA conference on Membership-Based Organizations of the Poor, Ahmedabad, India, January. Available at http://wiego.org/publications/waste-picker-cooperatives-developing-countries.Accessed 20 May 2012.

Medina, M. (2011).The informal sector – a driving force for recycling management. In *Recovering resources, creating opportunities: integrating the informal sector into solid waste management*, E. Gunsilius, S. Spies and S. García-Cortés (eds). Available at http://wiego.org/publications/recovering-.Accessed 18 December 2011.

Monni, S., R. Pipatti, A. Lehtilä, I. Savolainen, I and S. Syri (2006). *Global climate change mitigation scenarios for solid waste management.* Available at http://www.vtt.fi/inf/pdf/publications/2006/P603.pdf .Accessed 5 July 2013.

Myers, G. (2011). *African Cities Alternative Visions of Urban Theory and Practice*. London: Zed Books.

Nas, P.J.M. and R. Jaffe (2004). Informal Waste Management: Shifting the focus from problem to potential. *Environment, Development and Sustainability* **6**: 337 – 53.

Nigerian Newsworld (2011). Almajiri: The untold story. Available at http://www.nigeriannewsworld.com/content/almajiri-untold-story. Accessed 27 September 2012.

Nzeadibe, T.C. (2009). Solid waste reforms and informal recycling in Enugu urban area, Nigeria. *Habitat International* **33**: 93 – 9.

Onibokun, A.G. (1999). Synthesis and Recommendations. In *Managing the Monster: Urban Waste and Governance in Africa*, A.G. Onibokun (Ed.). Ottawa: IDRC.

Portes, A. and M. Castells (1989).*The Informal Economy: Studies in Advanced and Less Developed Countries*. Baltimore: Johns Hopkins University Press.

Rogerson, C.M. (2001). The Waste Sector and Informal Entrepreneurship in developing cities. *Urban Forum* **12**: 247 – 59.

Scheinberg, A. (2011). The informal sector in a modern, integrated, sustainable waste management system. In *Recovering resources, creating opportunities: integrating the informal sector into solid waste management*, E. Gunsilius, S. Spies and S. García-Cortés (eds). Available at http://wiego.org/publications/recovering-.Accessed 18 December 2011.

Sharp, J.P., P. Routledge, C. Philo and R. Paddison (2000).*Entanglements of Power Geographies of Domination /Resistance.* London and New York: Routledge.

Thorsen, D. (2012). *Children Working in the Urban Informal Economy: Evidence from West and Central Africa*. Briefing paper No. 3. Senegal: UNICEF.

Trager, L. (1987). A Re-Examination of the Urban Informal Sector in West Africa. *Canadian Journal of African Studies* **21**(2): 238 – 55.

Wilson, D.C, C. Velis and C.R. Cheeseman (2006).Role of informal sector in waste management in developing countries. *Habitat International* **30**: 797- 808

Velis, C.A., D.C. Wilson, O. Rocca, S.R. Smith, A. Mavropoulos and C.R. Cheeseman (2012). An analytical framework and tool ("InteRa") for integrating the informal recycling sector in waste and resource management systems in developing countries. *Waste Management and Research* **30**(43): 43 - 66.

Yusuf, O.S. (2011). A Theoretical Analysis of the Concept of Informal Economy and Informality in Developing Countries. *European Journal of Social Sciences* **20**: 624 – 36.

PART ONE
ENVIRONMENTAL SUSTAINABILITY

1

MATERIAL AND ENERGY-RECOVERY POTENTIAL OF MUNICIPAL SOLID WASTE IN NIGERIA

Chidozie Charles Nnaji

Introduction

Nigeria is a developing country grappling with many challenges, not the least of which is the menace of ever-increasing municipal solid waste. In the past two decades, solid waste in Nigeria has been increasing both in quantity and diversity. In addition, emerging waste streams, including e-waste and polythene/cellophane materials, are presenting an enormous challenge to already overwhelmed and incapacitated waste-management authorities. Individuals are often found in various Nigerian cities picking through this massive volume of garbage to recover materials of value. Material recovery is a precursor to the recycling and reuse of elements of municipal solid waste. In most developing countries where waste management constitutes a huge challenge, the volume of solid waste generated is usually reduced through material recovery. The key players are the waste pickers, usually referred to as scavengers, and waste dealers. This chapter discusses the material and energy recovery potential of municipal solid waste in Nigeria, with an emphasis on the role of waste pickers.

Status of Waste Storage and Collection in Nigeria

Waste storage and collection constitute a crucial stage in waste management. Proper waste storage makes for ease of collection, recovery and recycling. Waste storage has both environmental and health implications. Bins for waste are usually located within the house or in the corridors and harbour disease vectors such as houseflies, mosquitoes and cockroaches. Over time, these bins become heavily contaminated, transmitting germs upon the slightest contact with the skins of users. Most municipal waste management authorities encourage bagging of waste by providing polythene bags to households. Bagging makes waste collection less messy. Plastic bins, followed by polythene bags, are the commonest forms of waste storage container. Other, unconventional, waste receptacles include paper bags, used drums and sacks.

The choice of receptacle is usually based on the nature of waste to be stored, ease of conveyance, durability and affordability. Storage and collection form the final link between waste generators and waste managers, and can be used as a control. It is easier to sort waste at source than after collection. Nnaji and Utsev (2011) suggested that sorting at source (SAS) can be achieved by providing different waste bins for different categories of waste. Waste can be sorted into the following broad categories: biodegradable and non-biodegradable, compostable and non-compostable, recyclable and non-recyclable, combustible and non-combustible or into generic forms such as plastic, bottle/glass, ferrous metal, non-ferrous metal, paper, rubber and organic waste. This approach of course necessitates the use of several waste bins by households, but its advantages far outweigh this inconvenience. Studies have shown that some form of sorting at source is already being practised in Nigeria, though at a rudimentary level. Achi et al. (2012) observed that about 70 per cent of those who sort their waste at source in Abeokuta did so for monetary benefit. They also found some correlation between sorting and cultural background. While about 56 per cent of people in Abeokuta, southwest Nigeria, have tried one form of sorting or another, as Ukpong and Udofia (2011) have observed, no sorting at source was practised in Uyo, southern Nigeria. The present study concludes that SAS is a very important step towards effective recovery of clean materials for subsequent reuse and recycling and that this approach needs to be encouraged if recycling performance is to be improved in Nigeria.

Whenever the bins are filled, there arises the need for collection. However, studies have shown that about 30 to 60 per cent of solid waste generated in Nigeria is not collected (Benedine and Abbas 2011; Ogwueleka 2009). This can be attributed to (i) low accessibility resulting from poor road networks; (ii) technical, economic and management failures on the part of waste management authorities; (iii) inadequate facilities;(iv) refusal to pay collection fees by residents; and (v) non-optimisation of routes. Apart from the federal capital territory, Abuja, where about 56 per cent of residents are effectively served by the waste management authority, about 80 per cent of residents in other cities get rid of their waste by unauthorised open dumping, open air burning, burial and other unconventional means. Most municipal waste management authorities are overwhelmed by the enormous amount of waste generated, as well as the inadequacy and frequent breakdown of collection vehicles, low levels of manpower, poor data collection and management, poor wages and illiteracy. For instance, the Enugu State Waste Management Authority has only two collection trucks serving Nsukka metropolis and environs, with a population of about 97,000 residents. One of the trucks is currently out of service.

Refusal by residents to pay waste collection fees can be attributed to the poor services they get from waste management authorities. Besides, some people believe they should not pay for waste collection, because they have a right to that service for free. Because of the inability of waste management authorities to cope with the volume of waste generated, many Nigerian cities have designated dumps where households can take their waste. These dumps are visited in turn by collection vehicles to cart away the accumulated waste. Hence, waste management authorities do not offer services at the household level, but in cells. Some of these designated dumps are painfully far for some households, which, as a result, create open dumps close by. Such dumps are illegal and are therefore left unattended by authorities. Consequently, they keep growing in size and number. In most cases, collection vehicles travel along major city routes, collecting waste from the dumps located near to major roads. Some enterprising individuals have seized the opportunities presented by this situation to earn a decent living. Besides the waste pickers and middlemen, there are private firms that are involved in waste management. Sometimes, big institutions and municipalities contract waste management to these firms for an agreed fee. They can also collect waste from households at agreed intervals and fees. Compared with the municipal waste management authorities, these firms charge much higher fees. They are mostly patronised by the affluent living in highbrow areas of the cities, since the poor cannot afford the costs. These firms can afford expensive waste management equipment and their staffs are paid much better than government personnel.

Material and Energy Recovery Potential of Solid Waste in Nigeria

Material recovery is a precursor of the recycling and reuse of components of municipal solid waste. As noted at the outset of this chapter, in most developing countries where waste management constitutes a huge challenge, the volume of solid waste generated is usually reduced through material recovery. In this process, waste pickers and waste dealers play a key role. Some dealers go from house to house buying up reusable and recyclable waste materials, mostly in the form of bottles, plastics, metal tins and old footwear. They also visit auto repair shops to buy scrap metal, and dump sites to retrieve other valuables. Waste bought from homes or picked from dump sites is sold to scrap dealers, who serve as the middlemen between the waste pickers and the industries needing this waste as raw material. As shown in Figure1.1, metals and glass constitute close to 50 per cent of the waste materials sought by waste pickers because of their relatively high values. These items are sold by weight to middlemen who then resell them to industry (Figure 1.2). Figure 2 is a result of a survey undertaken by the author at Nsukka, Nigeria. Rust and deformation do not affect the price of these materials. Iron is melted and recycled by industry or sold to local blacksmiths for the fabrication of pots, spoons, plates, hoes and other farming implements. Copper is reused for the manufacture of copper wire, while aluminium is used to make roofing sheets. Bottles are either sold to bottling companies or to petty traders, who use them to package such edible items as honey, fried groundnut, roasted cashew nut, etc. Usually, broken bottles are not recovered by waste pickers. Plastic is used in the making of buckets, cups, spoons and plates, while paper is used for toilet rolls, shoe lining and in furniture making.

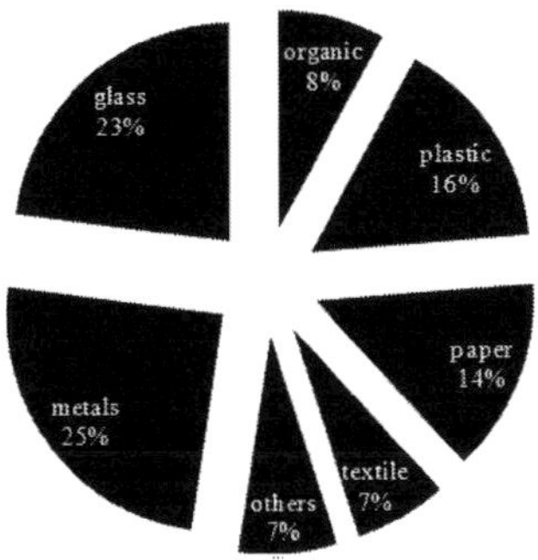

Figure 1.1:
Constituents of waste recovered in Maiduguri (Dauda and Osita 2003)

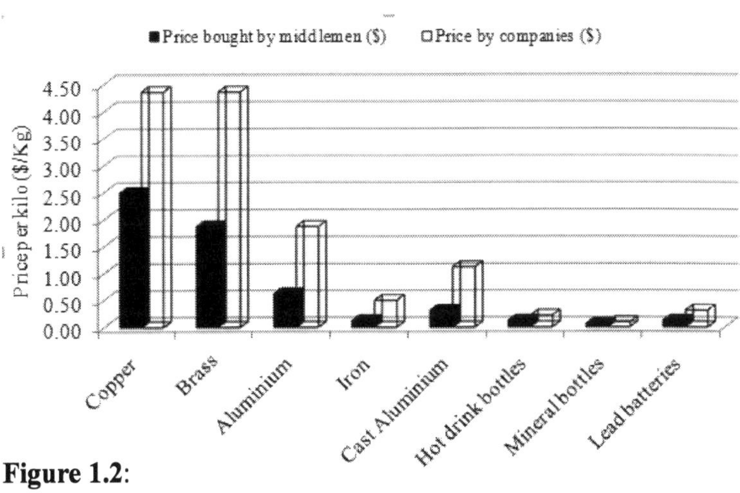

Figure 1.2:
Price per kilogram of recyclables (Author's fieldwork, 2012)

Waste picking has the double advantage of providing employment and also drastically reducing the volume of waste generated. Though waste picking is a lowly vocation, most of these individuals earn much more than the national minimum wage. Nzeadibe (2009) found that about 55 per cent of waste pickers are petty traders, with little or no education. Despite the crucial role waste pickers play in solid waste management in Nigeria, they also constitute a form of social and environmental hazard. Waste pickers have been implicated in a number of thefts. Moreover, in order to recover valuable non-combustible waste from a refuse dump, the waste pickers set the dump ablaze. After the waste heap has been consumed by fire, the waste pickers then go through the burnt remains to recover the metal scraps left behind. Uncontrolled burning of waste represents a serious source of air pollution. Due to the high moisture content of municipal solid waste, usually 20 per cent to 60 per cent, combustion is usually incomplete, giving rise to emissions of large amounts of carbon dioxide (CO_2) and smoke. Smoke resulting from the burning of plastics is known to cause chronic obstructive pulmonary disease (COPD), among other sicknesses. During the combustion of waste, greenhouse gases responsible for climate change are emitted. Besides, heat energy generated during uncontrolled combustion is lost to the environment. Sometimes, under-age children are seen at dumpsites picking out valuables. Waste pickers face the risk of cuts from broken bottles, nails and sharp metal objects, as well as burns from hot piles of waste. They also face the risk of being bitten by poisonous snakes while picking waste from dumps in bushy areas. Despite the tedious and demeaning nature of their work, waste pickers are often short-changed by middlemen, who sell the items for about twice the amount they pay the waste pickers (Figure 1.2). For instance, copper, brass and aluminium are bought from waste pickers at \$2.49, \$1.87 and \$0.62 respectively and sold by the middlemen at \$4.36, \$4.36 and \$1.87 respectively. Table 1.1 shows the uses to which scavenged materials may be put.

Table 1.1:
Uses of scavenged materials (Author's fieldwork, 2012)

	Material	Uses
1	Iron	Melted and cast into different items, such as rods, farm implements, etc.
2	Aluminium	Melted and cast into aluminium sheets, beverage cans, etc.
3	Brass	Melted and used in making musical instruments, etc.
4	Copper	Used in making copper wire, etc.
5	Bottles	Mostly reused in beverage industries. Used locally in packing edible nuts such ascashews, groundnuts, etc. for sale. Used for selling honey, red oil, local gin and local roots and herbs for human consumption.
6	Rubber	Used in making tyres, rubber slippers, etc.
7	Plastics	Re-usable plastics are melted and used to make cups, toys, buckets, spoons and plates
8	Papers/carton	Used to make shoe linings, furniture, toilet tissue and in paper mills
9	Polythene	Washed, chopped into flakes, dried, melted, pelletized and recycled

The main motivation for waste pickers in waste management is pecuniary, and hence their involvement is heavily skewed towards certain materials. Their activities are often carried out with gross disregard for the environment and in utter violation of environmental protection laws. Waste pickers can be harnessed for effective solid waste management through supervision. They can be made to register with municipal waste management authorities for effective monitoring. Moreover, their role can be enhanced by organising seminars and workshops for them. Waste pickers can also be employed for the organised and holistic sorting of solid waste at dump sites, instead of the current haphazard practices. The organic portion of municipal solid waste is of little or no value to waste pickers, because there is no market for it. Municipal waste management authorities can create the market by establishing composting plants. Lessons from Ondo State show that composting is a viable form of material recovery from municipal solid waste. Ondo State Integrated Waste Recycling and Treatment Plant (OSIWRP) is a $1.5 million project that currently employs about 80 workers. The plant mainly recycles plastics and produces organic fertiliser.

Regarding energy recovery from municipal solid waste, it is estimated that about 47 per cent of such solid waste in Nigeria is compostable and 32 per cent is recyclable, while 21 per cent is not of much value, consisting mainly of dust, ashes, broken bottles, diapers, paper, etc. A substantial proportion of non-compostable and non-recyclable waste items are not entirely useless. Some are combustible in a dry state and therefore can be a veritable source of energy. Energy can be recovered from solid waste by burning it in an incinerator, or by anaerobic digestion to produce biogas. Biogas, which is composed of about 70 per cent methane (CH_4) and 30 per cent carbon dioxide, can also be recovered from constructed landfills. Hence, besides material recovery, energy recovery is possible from municipal solid waste. Figures 1.3 and 1.4 show the energy potential of various components of municipal solid waste. It can be seen that polythene, polystyrene, tyre and other forms of plastic have the highest energy potential. Fortunately, the rate of generation of these materials has grown tremendously in the past few decades, and they have been found to comprise close to 20 per cent of the municipal solid waste generated in cities such as Maiduguri (Dauda and Osita 2013), Awka (Ifeoma et al. 2011), Gboko (Akpen and Aondoakaa 2009) and Ibadan business premises (Lade et al. 2012). Amber et al. (2012) observed that the heating value of mixed municipal solid waste is about one-third of the calorific value of coal. Using the calorific value of 2294.853KJ/kg given by Amber et al. (2012) and an average waste generation of 0.49kg/capita/day, the energy potential derivable from municipal solid waste per capita per day is approximately 2KJ. This value translates to a whopping 182GJ/day. If properly harnessed, this energy source can go a long way to alleviate the power problems currently confronting the nation.

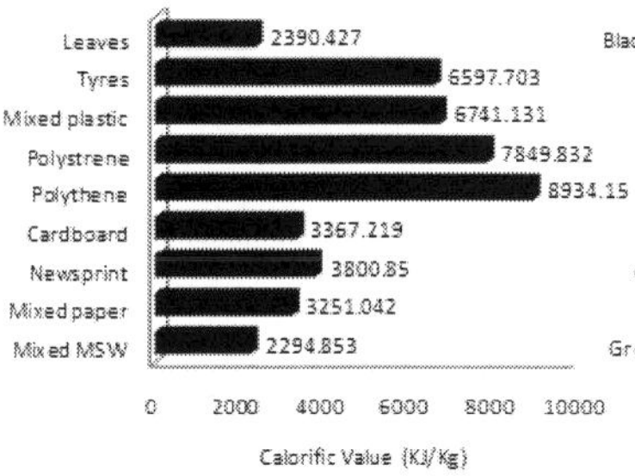

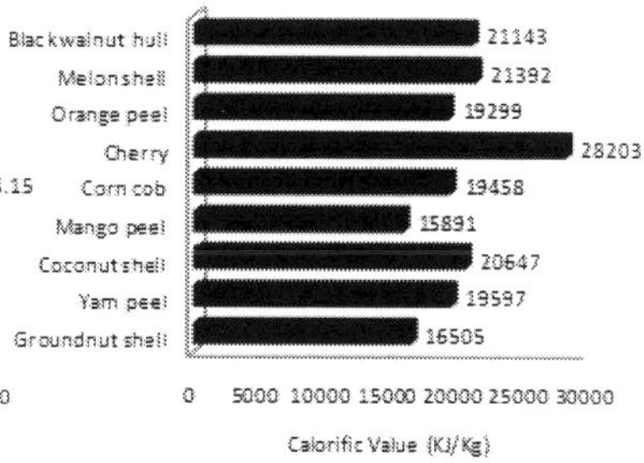

Figure 1.3:
Calorific value of combustible components of municipal solid waste (Amber, et al., 2012).

Figure 1.4:
Calorific values of plant-based organic components of municipal solid waste (Oumarou et al, 2012).

Despite the huge energy potential associated with municipal solid waste, the technicalities involved in harnessing such energy represent a formidable challenge for most developing countries. Amber et al. (2012) outlined the challenges facing energy recovery from waste in Nigeria as (i) lack of technology; (ii) lack of data on the specific composition of specific landfills; (iii) strength of solid waste management policy; (iv) lack of proper environmental education; (v) low level of awareness and sensitisation; and (vi) income status of individuals. The root of these challenges lies in the poor status of solid waste management in Nigeria. For instance, most of the disposal sites generally referred to as “landfills” are nothing but glorified open dumps. These disposal sites differ from constructed landfills in the following ways: (i) they are usually not designed or constructed; (ii) their location is arbitrary; (iii) there is usually no conformity with environmental protection laws; (iv) wastes are usually not sorted before disposal; (v) they are not lined to prevent groundwater contamination; (vi) there are no barriers to prevent overland flow of leached materials during the rainy season; (vii) there is no compaction of waste and no soil cover; and(viii) they are usually close to people’s houses. The few constructed landfills are very poorly maintained, and rapidly degenerate into open dumps. For this reason, no energy can be recovered from them. Besides the loss of potential energy is the consequence of unregulated biogas emissions into the atmosphere. Gases from dump sites and landfills are greenhouse gases, which contribute to global warming and climate change if they are not captured and properly utilised.

Conclusions

This chapter has examined the resource and energy potential of municipal solid wastes in Nigerian cities. It notes that material recovery is a precursor to recycling and reuse of components of municipal solid waste. The findings indicate there is huge potential for the recovery of the energy fraction of municipal solid waste and materials for recycling and reuse. In the chapter, it is argued that that it is waste pickers that are, in spite of their crude methods, presently driving the recycling system of Nigerian cities and that they are doing so mostly as a means to earn a living rather than for environmental protection. The argument is made that the entrepreneurial skills of waste pickers could be harnessed for effective solid waste management through the proper supervision of their activities. In this context, waste pickers can be made to register with municipal waste management authorities for effective monitoring.

Moreover, their role can be enhanced by organising seminars and workshops for them. Waste pickers can also be employed for the organised and holistic sorting of solid waste at dump sites, in place of the current haphazard activities. Public policy should therefore aim to empower these individuals and groups to improve their recycling efficiency and their lives in the face of the daunting health and social implications of their trade. This could go a long way towards reducing poverty among the waste pickers. Finally, the present study argues that SAS is a very important step towards effective recovery of clean materials for subsequent reuse and recycling. Adopting this approach will encourage improved recycling performance in Nigeria.

References

Abah, S.O. and E. Ohimain (2010). Assessment of dumpsite rehabilitation potential using integrated risk based approach: A case study of Eneka, Nigeria. *World Applied Sciences Journal* **8**(4): 436 – 42.

Achi, A.A., E.O. Adeofun, A.M.Gbadebo, G.C.Ufoegbunam and J.A. Oyedepo (2012).An assessment of solid waste management practices in Abeokuta, South West Nigeria. *Journal of Biology and Chemical Research* **29**(2): 177 – 88.

Adeoye, P.A., M.A. Sadeeq, J.J. Musa and E.A. Segun (2011). Solid waste management in Minna, North Central Nigeria: Present practices and future challenges. *Journal of Biodiversity and Environmental Sciences* **1**(6): 1 – 8.

Adewumi, I.K. M.O. Ogedengbe, J.A. Adepetu and Y.L. Fabiyi (2005).Planning Organic Fertilizer Industries for Municipal Solid Wastes Management, *Journal of Applied Sciences Research* **1**(3): 285 –91.

Afangideh, A.I., K.U. Joseph and J.E. Atu (2012).Attitude of urban dwellers to waste disposal and management in Calabar, Nigeria. *European Journal of Sustainable Development***1**:22-34

Afon, A. (2007). An analysis of solid waste generation in a traditional African city: The example of Ogbomosho, Nigeria. *Environment and Urbanization* **9**: 527 –37.

Akpen, G.D. and S.C. Aondoakaa (2009).Assessment of solid waste management in Gboko town.*Global Journal of Environmental Sciences* **8**(2): 71 –7.

Amber, I., D.M. Kulla and N. Gukop (2012).Municipal waste in Nigeria generation, characteristics and energy potential of solid. *Asian Journal of Engineering, Science and Technology* **2**(2): 84 –8.

Amber, I., D.M. Kullaand N. Gukop (2012). Generation, characteristics and energy potential of solid municipal waste in Nigeria, *Journal of Energy in Southern Africa* **23**(3): 47 – 51.

Audu, G.B., G. Bogoro and S.M. Nghalmi (2013). Potentials and constraints of household solid waste segregation in Bauchi Metropolis. *International Research Journal of Arts and Social Sciences* **2**(4): 99 – 116.

Babalola, A., T.H. Ishaku, I. Busu and M.S. Madid (2010). The practice and challenges of municipal solid waste management in Damaturu, Yobe State, Nigeria. *Journal of Environmental Protection* **1**: 348 – 88.

Babayemi, J.O. and K.T Dauda (2009). Evaluation of solid waste generation, categories and disposal options in developing countries: A case study of Nigeria. *Journal of Applied Sciences and Environmental Management* **13**(3): 83 – 8.

Benedine, T.A. and I.I. Abbas (2011). The impact of spatial distribution of solid waste dumps on infrastructure in Samaru, Zaria, Kaduna State, Nigeria using geographic information systems (GIS).*Research Journal of Information Technology* **3**(3): 113 – 17.

Bichi, M.H. and D.A. Amatobi (2013). Characterization of household solid wastes generated in suburban areas of Kano in Northern Nigeria. *American Journal of Research Communication* **1**(4): 165 – 71.

Dauda, M. and O.O. Osita (2003). Solid waste management and re-use in Maiduguri, Nigeria. 29th WEDC International Conference: Towards the Millennium Development Goals, Abuja, Nigeria, pp. 20 –3.

Egbere, O.J., J.U. Utelim and F.O. Opiah (2001). Municipal solid waste segregation and their health hazard implications in Angwan Rogo of Jos, Plateau State, Nigeria. *Journal of Environmental Sciences* **4**(1): 1 – 4.

Egun, N.K. (2012). The waste to wealth concept: waste market in Delta State, Nigeria. *Greener Journal of Social Science* **2**(6): 206 – 12.

Etusim, P.E., D.C. Amuzie, S.N. Umeham, P.Ogwo and A.K. Obialor (2013). Twelve day study on characterization of waste generated at an urban market in Okigwe, Imo State Nigeria. *Journal of Physical Sciences and Environmental Safety* **3**(1): 17 - 23.

Fakere, A., G. Fadair and O. Oriye (2012). Domestic Waste Management and Urban Residential Environment: Focus on Akure,

Nigeria. *International Journal of Engineering and Technology* **2**(5): 878 – 87.

Ibrahim, S.I., A.A. Aworonti and L.T. Ajibade (2012).Storage and treatment system of solid waste as strategy for sustainable development in Illorin Metropolis, Nigeria. *Advances in Applied Science Research* **3**(2): 801 –8.

Ifeoma, M., O.U. U, E. Nkiru, O.C. N, A.N. E. (2011). Public Health Implication Of Household Solid Waste Management In Awka South East Nigeria. *Internet Journal of Public Health* **1**(1). DOI: 10.5580/265d.

Igbinomwanhia, D.I. (2011). Status of Waste Management. In *Integrated Waste Management*. Volume II, S. Kumar (ed). Intech, pp.12 – 34.DOI: 10.5772/935.

Lade, O., O. Agbede and O. Ilori (2012). Management of Solid Waste in a Market: Case Study of Bodija Market, Ibadan. *Nigeria Journal of Environmental Science and Engineering* **1:** 845 – 50.

Lawal, A.M. and I. Garba (2013).Study of the Energy Potential of Solid Waste in Bauchi Town. *International Journal of Computational Engineering Research* **3**(5): 1 – 7.

Mohammed, A.S. and G.O. Chukwuma (2011).Strategy for Remediating the Impacts of Solid Wastes on Soil and Groundwater Quality in Minna, Nigeria. *Journal of Innovative Research in Engineering and Science* **2**(3): 173–84.

Nnaji, C.C. and T.T. Utsev (2011). Climate Change and Solid Waste Management: A Balanced Assessment. *Journal of Sustainable Development in Africa* **13**(7): 17 – 34.

Nzeadibe, T.C. (2009). Development drivers of waste recycling in Nsukka Urban Area, Southeastern Nigeria.*Theoretical and Empirical Researches in Urban Management* **3**(12); 137 – 49.

Ogwueleka, T.C. (2009).Municipal solid waste characteristics and management in Nigeria. *Iranian Journal of Environmental Health Science and Engineering* **6**(3): 173 –80.

Onwughara, N.I., I.C. Nnorom and O.C. Kanno (2010).Issues of roadside disposal habitat of municipal waste, environmental impacts and implementation of sound management practices in developing country Nigeria. *International Journal of Environmental Science and Development* **1**(5): 409 – 18.

Oumarou, M.B., M. Dauda, A. T. Abdulrahim and, A.B. Abubakar (2012).Characterization and generation of municipal solid waste in North Central Nigeria. *International Journal of Modern Engineering Research* **2**(5): 3669 – 72.

Oyelola, O.T. and A. Babatunde (2012).Characterization of domestic and market solid wastes at source in Lagos metropolis, Lagos, Nigeria. *African Journal of Environmental Science and Technology* **3**(12): 430 – 7.

Salam, M.A. L.M. Hossain, S.R. Das, R. Wahab and M.K. Hossain (2012).Generation and assessing the composition of household solid waste in commercial capital city of Bangladesh. *International Journal of Environmental Science, Management and Engineering Research* **1**(4): 160 – 71.

Sha'Ato, R., S.Y. Aboho, F.O. Oketunde, I.S. Eneji, G.Unazi, and S.Agwa (2007).Survey of solid waste generation and composition in a rapidly growing urban area in Central Nigeria. *Waste Management* **27**: 352 – 8.

Ukoje, J.E. (2011). Analysis of the determinants of participation of stakeholders in solid waste management in Zaria, Nigeria. Ph.D thesis submitted to the School of Postgraduate Studies, Ahmadu Bello University, Zaria, Nigeria.

Ukpong, I.E. and E.P. Udofia (2011).Domestic Solid Waste Management in a Rapidly Growing Nigerian City of Uyo. *Journal of Human Ecology* **36**(3): 229 – 35.

2

EXTENDED PRODUCER RESPONSIBILITY AS A TOOL FOR MANAGING E-WASTE IN SUB-SAHARAN AFRICA: KEY ISSUES

Chukwunonye Ezeah and Jak A. Fazakerley

Introduction

Electronic waste (e-waste) or Waste Electrical and Electronic Equipment (WEEE) are two terms often used to describe End-of-Life (EoL) electronic devices that the consumer has discarded or no longer sees to be of value (Jang 2010). Such devices include laptop computers, mobile phones, video cameras, televisions, hair dryers, irons, cookers and washing machines (Robinson 2009). For the purpose of this article, the two terms will be used interchangeably. E-waste is non-homogenous and complex in terms of material, chemical and polymer composition, and of mechanical properties (Gramatyka et al. 2007). Such waste not only varies between equipment types, but also between appliances within a single equipment type (Chancerel and Rotter 2009). Widmer et al. (2005) argue that given the diverse range of materials found in WEEE; even a generalised listing of the material composition of the entire waste stream is difficult. WEEE contains a vast array of precious metals (palladium, platinum, gold, silver), base metals (lead, nickel, zinc, copper, aluminium) and critical metals (antimony, cobalt, gallium, terbium, europium, indium, dysprosium) (Hidy et al. 2011). These metals are clearly valuable, and are in fact the primary economic incentive for recycling e-waste (Tuncuk et al. 2012). Toxic substances, including metals (lead, mercury, arsenic, cadmium, selenium and chromium) and organic chemicals (flame retardants, Polychlorinated Biphenyls – PCBs – and polyvinyl chloride) are also used to manufacture hazardous components and materials for electrical products (Cui and Forssberg 2003). If not handled, treated and disposed of appropriately, they can form volatile metal halides or produce toxic organic compounds, such as dioxins and furans, which can result in environmental contamination and have negative effects on human health (Jang 2010).

Electrical and electronic equipment is continually being developed, applied and consumed in very large quantities worldwide (He et al. 2006). Wath et al. (2006) believe that the electronics industry is the world's largest and fastest growing manufacturing sector. This situation has been partly attributed to the increasing affordability of electronic equipment, rapid innovation in technology and design, market expansion, increased consumer demand for newer electronic products, global population growth, rapid GDP growth, short innovation cycles, and the fact that global markets are far from saturation point (Nnorom and Osibanjo 2008a; Rahimifard et al. 2009). However, the fast replacement process means that large numbers of appliances are becoming obsolete, despite the fact they are often still quite functional. UNEP (2009) calculated the global production of e-waste to be between 20-50 million tonnes per year. A large amount of this is generated in countries of the OECD (Organisation for Economic Co-operation and Development) such as Europe, the US and Australia (Widmer et al. 2005). Though per capita e-waste generation in sub-Saharan Africa is relatively small by comparison, it is expected to increase rapidly due to increased penetration rates and a decrease in product life spans (Manomaivibool 2009). This view is reinforced by Oliveira et al. (2012), who note how the sales of electronic products in sub-Saharan Africa are predicted to rise sharply in the coming decade, with sales of products such as mobile phones already having increased by a factor of 100 over the past decade.

Increasingly, improper management of e-waste in sub-Saharan Africa is becoming a subject of great concern to industry watchers and society alike. The situation is particularly worrisome in countries such as Nigeria, Kenya and Ghana, where environmental and health hazards have resulted from a combination of low-end management alternatives such as conventional disposal in landfills, open burning and backyard recycling (Manomaivibool 2009). It is therefore important to innovate, develop and adopt robust environmental management systems that incorporate regulations, policies, legislation, guidelines, strategies, infrastructure and technology so as to identify and manage these impacts in a more sustainable manner (DEFRA 2006).

In contrast to the current e-waste situation in sub-Saharan Africa, OECD member states understand the need for a very robust regulatory framework to tackle the e-waste problem and have put in place a very comprehensive legislative infrastructure to tackle the negative impacts of e-waste disposal (Wath et al. 2010). One approach they have adopted is an Extended Producer Responsibility (EPR) programme. EPR is an environmental policy principle in which the responsibility of those who manufacture or import products within a particular jurisdiction is extended to the post-consumer stage of the product's lifecycle. This lifecycle perspective assigns responsibility for all the activities, ranging from raw material extraction, manufacturing, distribution, use, through to reuse, recycling, recovery and disposal (Oliveira et al. 2012). If the initiative is well thought out and adequately implemented, it has the potential to divert and recover considerable amounts of waste (as is the case in Switzerland).

It is difficult to transfer the e-waste management system in OECD countries to non-OECD countries on account of many country-specific issues, such as socioeconomic conditions; lack of control and organisation; severe infrastructure shortfalls for the collection, treatment and disposal of WEEE; absence of WEEE-specific legislation; and a lack of commitment (Oliveira et al. 2012; Li et al. 2013). Furthermore, in non-OECD countries the structure of product markets is very different, with a large share of non-identifiable producers, historical orphan products, large black markets for some electronic products and ever increasing illegal imports of WEEE (Manomaivibool 2009). As such, implementing an efficient and effective EPR policy in developing countries such as Nigeria is bound to be both complex and challenging.

E-waste policies and handling in sub-Saharan Africa

Imported e-waste

Currently the main disposal route for e-waste in most OECD countries is through export to developing and transitional countries such as Pakistan, Vietnam, China, India and sub-Saharan Africa (Nigeria, Ghana, Kenya, South Africa) (Nnorom and Osibanjo 2008b; UNEP 2009). Some of the products are "donated" to the economically and technologically weak (hospitals, orphanages, schools) by official consumers and well-known charitable organisations in the name of "helping the poor" and "bridging the digital divide" (Oswald and Reller 2011). Zaman and Lehmann (2011) emphasise how electronic waste in high-consuming countries is often portrayed as a valuable and useful resource in low-consuming countries. However, large quantities are shipped illegally, often mislabelled as "reusable equipment" and "mixed metal scrap" (Wi Paschal dmer et al. 2005; Nnorom and Osibanjo 2008b; Oswald and Reller 2011). *Ab initio*, this approach is primarily driven by economic considerations, that is low labour costs, extenuated trade, a large demand for low-priced secondary materials and the cost of managing waste in an environmentally sound manner (Kahhat et al. 2008; Hidy et al. 2011; Ongondo et al. 2011). Nnorom and Osibanjo (2008b) estimate that between 25 and 75 per cent of e-waste exports are unusable (non-functional or unrepairable) junk. Due to such fictitious hustle, it is becoming increasing difficult to distinguish between products destined for reuse and refurbishment and e-waste destined for recovery or final disposal (Oliveira et al. 2012). As a result, exact statistics on the level of e-waste being managed in sub-Saharan Africa are unavailable.

The Basel Convention (1992) is the international legislation governing the management of e-waste. Its primary aims are to protect the health of humans and the environment against the adverse effects of the generation, storage, [transboundary] movement and disposal of EoL products (Osibanjo and Nnorom 2008b). Despite this, Nnorom and Osibanjo (2008b) indicate there is an increase in the transboundary movement of e-waste from developed to developing countries. Oliveira et al. (2012) believe that the classification systems used for e-waste streams is piecemeal and ad hoc, allowing for different interpretations and thus illegal transboundary movements. Furthermore, Kahhat et al. (2008) state that those countries that have not ratified the convention (such as the US) continue to export on the grounds that they are not violating international laws. As such, rigorous enforcement of the convention is needed as an alternative to implementation of a blanket ban on all imports (Manomaivibool 2009). Osibanjo and Nnorom (2008b) argue that such blanket bans (see Table 2.1) on imports of EoL electronics will massively reduce employment opportunities in the informal sector.

Table 2.1:
Legislation and regulations related to imports and exports of WEEE and Used Electrical and Electronic Equipment (UEEE) in sub-Saharan Africa.

Country	Import of WEEE	Import of UEEE	Legislation/ Regulation
Nigeria	Banned (2011)	Permitted	Guide for Importers of UEEE into Nigeria
Ghana	Permitted (2011)	Permitted	Ghana e-waste Country Assessment
Kenya	Not Defined	Not Defined	Not Defined

South Africa	Not Defined	Not Defined	Not Defined
Uganda	Not Defined	Not Defined	Not Defined
Senegal	Not Defined	Not Defined	Not Defined
Liberia	Not Defined	Not Defined	Not Defined
Benin	Not Defined	Not Defined	Not Defined

Source: Adapted from Li et al. (2013).

Informal sector

Historically, the informal sector in developing countries has almost always been framed by researchers within a poverty perspective. Beyond poverty, however, informal sector activities could be viewed from a social innovation as well as an entrepreneurial perspective (Ezeah et al. 2013). E-waste recycling is a lucrative business in sub-Saharan Africa and is dominated by the informal sector (Li et al. 2013). The sector consists of a network of collectors, dealers, dismantlers and recyclers, all of whom add value at each point of the chain. The potential economic value of some of the metal-based products (e.g., gold, copper, silver) makes recovery an attractive proposition and so plenty of job opportunities open up for those who are unable to find work elsewhere (Manomaivibool 2009). Such activity can therefore be described as market driven. With few if any material recovery facilities or technologies for e-waste recycling in place (Li et al. 2013), the products are often processed in backyards or small workshops using labour-intensive, very risky and primitive tools, methods and technologies, for instance chemical leaching, cracking and open burning, to recover the valuable precious metals (He et al. 2006; Osibanjo and Nnorom 2008a). The rest is disposed of in open dumps, unlined landfills or surface water bodies (Oswald and Reller 2011; Ongondo et al. 2011). This results in ineffective resource exploitation and the use of scarce land resources as landfills (Nnorom and Osibanjo 2008a). Furthermore, Oliveira et al. (2012) believe that backyard recycling is problematic for human health, as there are few if any safeguards. The workers often do not have protective clothing or equipment and are exposed to occupational hazards (Oswald and Reller 2011). Ecosystem health is also seriously undermined, as extreme localised pollution is followed by the uncontrolled migration of contaminants into food chains and water bodies (He et al. 2006). This can induce genetic mutation, respiratory tract infections and cytogenetic damage within the immediate population (Robinson 2009). BAN (2005) argues that, in effect, the population are forced to choose between "poverty and poison."

Manomaivibool (2009) believes that those in the informal sector do not wish to pursue formalisation of their work, as its perceived costs would outweigh the benefits. For example, informality allows for savings in terms of pollution abatement, worker health and safety measures and taxes. Furthermore, foreign companies are not convinced about investing in the region, despite low labour costs and ever increasing, readily available quantities of secondary resources. Oswald and Reller (2011) find this alarming, stating that the value of recovered resources actually exceeds recycling costs. This needs to be understood and addressed when framing EPR systems for sub-Saharan Africa (Wath et al. 2010).

Extended Producer Responsibility: Theoretical Concept

EPR is a waste management system created on environmental, economic and social principles (Widmer et al. 2005; He et al. 2006). The OECD (2001) describes it as "an environmental policy approach in which a producer's responsibility for a product is extended to the post-consumer stage of the product's life cycle, including its final disposal." This basically shifts responsibility upstream to the industries, importers, etc. and away from the municipalities (Widmer et al. 2005; Nnorom and Osibanjo 2008a). The producer is responsible for the take-back of the product, that is the collection, transporting, sorting, refurbishment, dismantling, shredding, treatment of recyclables/hazardous components, and landfill or incineration (with heat recovery) of the remaining waste (He et al. 2006). Effective EPR programmes should: promote clean production concepts and the eco-design of products and packaging so as to improve reusability and recyclability (i.e., "design for disassembly") (Jang 2010);

conserve natural resources and promote energy efficiency by ensuring high utilisation of the product and material quality (Nnorom and Osibanjo 2008a); promote data and enforcement mechanisms (DEFRA 2006); encourage technological and scientific innovation in the recovery sector (Nnorom and Osibanjo 2008c); prevent the emergence of new orphan products and free riders (Manomaivibool 2009); facilitate recycling and reduce third-party involvement (Oswald and Reller 2011); raise public awareness and change attitudes regarding appropriate handling, source separation and proper disposal through publicity and education (Khetriwal et al. 2009); provide consumers with incentives to return used products (Nnorom and Osibanjo 2008b); distribute costs in a way that is fair and just (Manomaivibool 2009); and promote the reuse market (Jang 2010). In short, the objective of EPR is to design upstream and downstream product systems that are economically sound, environmentally friendly and socially desirable.

Designing and characterising an EPR management system with clear and defined roles is essential for all actors. First of all, administrative, economic and informative policy instruments must be initiated (Table 2.2) and then a number of key parameters and approaches need to be carefully considered. These include legal regulation, system coverage, financing, and producer responsibility (Widmer et al. 2005).

Table 2.2:
Policy Instruments and Objective Solutions

Policy Instrument	**Objective Solutions**
Product take-back programmes	Mandatory take-back, voluntary take-back, trade-in scheme
Economic	Advanced recycling fee, deposit refund scheme, pre-disposal fee, material taxes
Voluntary industry practices	Voluntary codes of practice, public/private partnerships

Regulatory approaches	Disposal bans, minimum product standards, mandated recycling, restriction of certain components/materials

Source: Adapted from Nnorom and Osibanjo (2008b)

EPR initiatives in sub-Saharan Africa

Europe so far leads the way in framing and implementing policies to manage its e-waste stream (Khetriwal et al. 2009). More specifically, the Swiss are credited with establishing the first comprehensive e-waste management system from collection to disposal (Sthiannopkao and Wong 2012). However, Li et al. (2013) state that an EPR management model selected for one country cannot always be applied completely to another on account of various internal and external factors. As such, EPR models should be carefully analysed and take into account experiences from other countries. The overall target in planning a national EPR model must be the creation of a sustainable recycling society. It is not sufficient to confine responsibility only to producers. Retailers, governments, local authorities, the recycling industry (including the informal sector), NGOs and consumers all have specific roles in ensuring the total recycling of e-waste can happen in a way that is cost-efficient, sustainable, accessible and fair (Tanskanen 2013).

Financial sustainability of the system

There are many different models being applied globally for financing the particular costs in the recycling chain. In order for an efficient, self-sustaining and smoothly functioning system to be achieved, cost-sharing has to be agreed among all relevant actors. This includes trade associations, individual companies, recyclers, governmental agencies, NGOs, development agencies and consumers.

Producers' voluntary take-back scheme

This is described as "a marketing strategy based on an environmental ground and driven mainly by external influences" (Manomaivibool

2009: 142). Multinational corporations are under increasing pressure from civil society and global policymakers to take responsibility for the product's entire lifecycle. As such, a take-back scheme should be introduced even if no mandatory e-waste programme exists within the destination country. The system can be financed in a number of ways, such as by payment of an Advance Recycling Fee (ARF) or a Pre-Disposal Fee (PDF). From a psychological point of view, a consumer may be more willing to pay a small fee at the time of purchase than for the disposal of a broken or unwanted product (Wath et al. 2010). In addition, an ARF ensures that the product is collected, transported, decontaminated, recovered and recycled in an environmentally friendly way, rather than being dumped into the municipal solid waste stream or otherwise. By contrast, Khetriwal et al. (2009) argue that an advanced fee punishes rather than promotes products with a "green" design. Alternatively, a deposit-refund system could be implemented. In this case, consumers pay a fee when they purchase a product that is refunded when they return it to a certified recycler (Wath et al. 2010). This provides a financial incentive to return the product in an authorised manner. For this to work efficiently and effectively, records of sales must be kept up to date, thus adding to the administrative pressures on producers, retailers and recyclers.

Retailer trade-in scheme

Tax credits or discounts are given to consumers when they return their WEEE and opt to buy a new product from the same retailer (Manomaivibool 2009). Trade-in is a way to stimulate sales of new products by influencing a consumer's replacement decisions. This scenario can therefore be described as a market strategy driven by profit. Consumers are financially motivated and so may be willing to participate. However, for the system to work, uniform guidelines must be put in place that helps to identify manufacturers, producers and recyclers. Once identified, they are assigned with financial and administrative responsibilities to enhance the recycling chain (Wath et al. 2010). Manomaivibool (2009) believes that the fate of traded-in products, as well as rates of discounts offered, must also be defined to reduce the likelihood of black market diversion.

The two take-back options available highlight the need for a well-defined and organised system. Special logistical requirements for collecting e-waste from the source of its generation and transporting to the site of refurbishment, recycling, recovery and disposal are therefore essential. It is also important to note that an effective EPR programme is partly conditioned by the consumer's context. Wagner (2013: 501) argues that "there is a need to better understand the steps involved in participation so as to identify the specific elements that affect convenience." Very often, the convenience of a recycling programme is the most important non-socioeconomic determinant of whether an individual will participate. People often encounter problems relating to the proximity to the collection site and the ease of the process (i.e., sorting, segregating and storing e-waste) (Wagner 2013). The more inconvenient each step is, the lower the likely participation rate and thus the lower the diversion recovery rate. A primary goal should be increasing the convenience of the entire process.

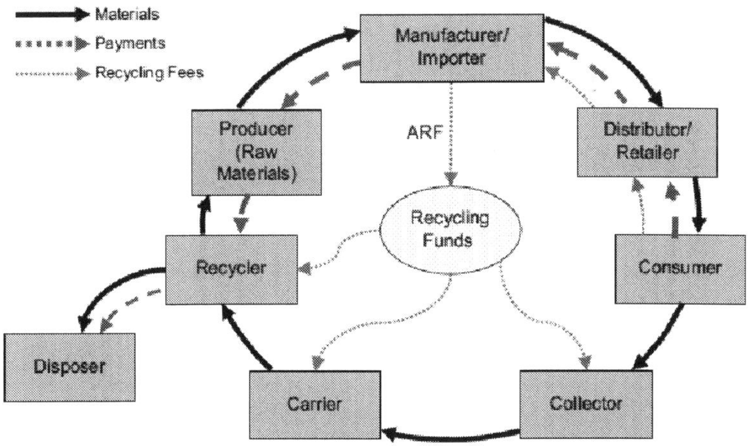

Figure 2.1:
Flow of materials and finances in the Swiss e-waste management system
Source: Khetriwal et al. 2009

Collective v. individual take-back

Individual take-backs are brand-specific, accepting only products from a particular producer. A collective take-back system, on the other hand, does not distinguish between brands of product types. Khetriwal et al. (2009) believe that all-inclusive take-back systems are more consumer-oriented by making it easier for consumers to drop their products off in one place (see Figure 2.1). Individual systems can be inconvenient for a consumer who has to dispose of different brands at different locations. This would also reduce efficiency for recyclers, who have to separate the appliances accordingly (Raymond 2002). This view is reinforced by Tanskanen (2013), who believes that the setting up of collection systems is most effective when organised in a collective manner. However, Nnorom and Osibanjo (2008b) argue that in order to promote design improvement, individual responsibility is needed. Many EU-based firms, including Sony, HP and IBM, support individual responsibility, as it gives them control of system financing and the implementation of meaningful targets (He et al. 2006).

Rahimifard et al. (2009) believe that because of this, many manufacturers have opted to subcontract the recovery and recycling of their products to the recovery industry (producer compliance scheme), which has very little knowledge of the products being recovered. However, Wath et al. (2010) argue that by devising support systems (i.e., infrastructure development), compliance costs can be brought down. Khetriwal et al. (2009) note how the introduction of producer responsibility organisations (PRO), such as Swiss Association for the Information, Communication and Organisational Technologies (SWICO) and Swiss Foundation for Waste Management (SENSE) in Switzerland, has led to a sustainable e-waste management system in terms of financing, collection, transportation and control systems. Very often, it is these not-for-profit organisations that "take the first step" in terms of operational responsibility. Khetriwal et al. (2009) strongly believe that "given the global footprint of the EEE industry, it would be possible to easily transfer operational and system knowledge gained from setting up such voluntary systems in Switzerland and apply them in other geographies." Having a working system in place before policy is introduced means that procedures can develop. Furthermore, if implemented efficiently and equitably, such an approach could create a convenient and transparent system and a level playing field for all actors involved. It would also overcome the issue of historical and orphan products as Advanced Recycling Fee (ARF) act as intergenerational contracts and so PROs are able to finance recycling irrespective of brand or date of manufacture (Wath et al. 2010).

The smooth functioning of the system can be ensured by transparent contract procedures and independent control and monitoring of material and financial flows at every stage (Khetriwalet al. 2009). This will help to identify free riders and ensure environmental standards are maintained. Sthiannopkao and Wong (2012) suggest that an e-steward initiative should be introduced for auditing and certifying recyclers and take-back programmes so that customers know which organisation is reliable and meets appropriate standards.

Key considerations

- Accurate and reliable data is urgently needed to estimate e-waste stocks in both upstream and downstream sectors, as well as the illegal trading in e-waste in sub-Saharan Africa. From this an inventory on e-waste can be produced.
- Consultation meetings should be held to assess e-waste at national and sub-regional levels; share experiences and best practices; and discuss strategies for promoting environmentally sound management options for waste.
- Careful design is required for the EPR programme, which should utilise appropriate policy instruments, monitoring and reporting mechanisms, financial assistance and infrastructure at different levels.
- The revolutionary work done in Europe cannot be directly transferred to sub-Saharan Africa without a solid understanding of the economic, social, environmental, temporal, spatial, situational and philosophical context of the area.
- The manner in which each state or region incorporates EPR into its legislation will differ in terms of scope, range, type of producer responsibility and funding mechanisms.

- The regulations in each country vary widely, so a regional-level guidance document distinguishing UEEE from WEEE is necessary for the control of illegal trafficking.
- An EPR programme can be a driver for formalising the informal recycling sector. Upon assuming physical responsibility, producers can only send collected e-waste to the formal sector. Informal sector organisations/cooperatives should therefore be developed that will lead to the integration of the sector into municipal solid waste management systems. They need to be supported by industry, local authorities, private partnerships and the general public. They should also be initiated in a way that does not neglect the environment.
- Integrating the informal sector into formal systems could significantly reduce e-waste collection costs and hence the recoverable value may exceed the collection costs. Thus an additional recycling fee is unnecessary.
- Producers should engage in research and development, especially with respect to greener designs (e.g., "design for recycling" and "design for disassembly") and hazardous material control. Producers who have control over the design of products and product systems are almost always at the centre of environmental improvements.
- Producers should form task force groups for communication and information, finance, legal and technical assistance.
- Well-defined roles and responsibilities for stakeholders will ensure a high level of compliance.
- Governments should introduce policy and legislation specifically dealing with e-waste. This must take into consideration the existing socio-cultural-economic background and practices.
- Import subsidies, tax credits and other incentives should be introduced to encourage investors and small and medium enterprises interested in the material recovery of e-waste. This will also ensure that monopolies do not thrive.
- All products need to be taken back, regardless of value, demand or manufacturer.

References

BAN (2005). The Digital Dump: Exporting Re-use and Abuse to Africa. Basel Action Network. 24 October. Jim Puckett (ed.). Available at: www.ban.org. Accessed on 20 April 2013.

Chancerel, P. and S. Rotter (2009).Recycling-oriented Characterization of Small Waste Electrical and Electronic Equipment. *Waste Management* **29:** 2336 – 52.

Cui, J. and E. Forsberg (2003). Mechanical Recycling of Waste Electric and Electronic Equipment: A Review. *Journal of Hazardous Materials* **99:** 243 – 63.

Department for Environment, Food and Rural Affairs (DEFRA) (2006).*Guidance on Best Available Treatment Recovery and Recycling Techniques (BATRRT) and Treatment of Waste Electrical and Electronic Equipment (WEEE)*. Available at: http://archive.defra.gov.uk/environment/waste/producer/electrical/documents/weee-batrrt-guidance.pdf. Accessed on 2 November 2012.

Ezeah, C., J.A. Fazakerley and C.L. Roberts (2013).Emerging trends in informal sector recycling in developing and transition countries. *Waste Management* **33**(11): 2509 – 19. DOI: http://dx.doi.org/10.1016/j.wasman. 2013.06.020

Gramatyka, P., R. Nowosielski and P. Sakiewicz (2007).Recycling of Waste Electrical and Electronic Equipment. *Journal of Achievements in Materials and Manufacturing Engineering* **20:** 535 – 8.

He, W., G. Li, X. Ma, H. Wang, J. Huang, M. Xu and C. Huang (2006). WEEE Recovery Strategies and the WEEE Treatment Status in China. *Journal of Hazardous Materials* **136:** 502 – 12.

Hidy, G.M., W. Alcorn, R. Clarke, D. Smith and V. Thomas (2011). Environmental Issues and Management Strategies for Waste Electronic and Electrical Equipment. *Journal of the Air and Waste Management Association* **61:** 990 – 4.

Jang, Y. (2010) Waste Electrical and Electronic Equipment (WEEE). Management in Korea: Generation, Collection, and Recycling Systems. *Journal of Material Cycles and Waste Management* **12:** 283 – 94.

Kahhat, R., J. Kim, M. Xu, B. Allenby, E. Williams and P. Zhang (2008).Exploring E-waste Management Systems in the United States. *Resources, Conservation and Recycling* **52:** 955 – 64.

Khetriwal, D.S., P. Krawuchi and R. Widmer (2009). Producer Responsibility for E-waste Management: Key Issues for Consideration – Learning from the Swish Experience. *Journal of Environmental Management* **90:** 153 – 65.

Li, J., B.N. Lopez, I. Liu, N. Zhao, K. Yu and L. Zheng (2013). Regional or Global WEEE Recycling. Where to go? *Waste Management* **33:** 923 – 34.

Manomaivibool, P. (2009). Extended Producer Responsibility in a non-OECD Context: The Management of Waste Electrical and Electronic Equipment in India. *Resources, Conservation and Recycling* **53:** 136 – 44.

Nnorom, I.C. and O. Osibanjo (2008a). Electronic Waste (e-waste): Material Flows and Management Practices in Nigeria. *Waste Management* **28:** 1472 – 79.

Nnorom, I.C. and O. Osibanjo (2008b).Overview of Electronic Waste (e-waste) Management Practices and Legislations, and their Poor Applications in the Developing Countries.*Resources, Conservation and Recycling* **52:** 843 – 58.

Nnorom, I.C. and O. Osibanjo (2008c). Sound Management of Brominated Flame Retarded (BFR) Plastics from Electronic Wastes: State of the Art and Options in Nigeria. *Resource, Conservation and Recycling* **52:** 1362 – 72.

OECD (2001).*Extended Producer Responsibility: A Guidance Manual for Governments*. Paris: OECD.

Oliveira, C.R.D., A.M. Bernardes and A.E. Gerbase (2012). Collection and Recycling of Electronic Scrap: A Worldwide Overview and Comparison with the Brazilian Situation. *Waste Management* **32:** 1592 – 1610.

Ongondo, F.O., I.D. Williams and T.J. Cherrett,(2011). How Are WEEE Doing? A Global Review of the Management of Electrical and Electronic Wastes. *Waste Management* **31:** 714 – 30.

Osibanjo, O. and I.C. Nnorom (2008). Material Flows of Mobile Phones and Accessories in Nigeria: Environmental Implications and Sound End-of-Life Management Options. *Environmental Impact Assessment Review* **28:** 198 – 213.

Oswald, I. and A. Reller (2011). E-Waste: A Story of Trashing, Trading, and Valuable Resources. *GAIA – Ecological Perspectives for Science and Society* **20:** 41– 7.

Rahimifard, S., M.S. Abu-Bakar and D.J. Williams (2009).Recycling Process Planning for the End-of-Life Management of Waste from Electrical and Electronic Equipment.*CIRP Annals – Manufacturing Technology* **58:** 5 – 8.

Raymond, M. (2002).Getting Your Product Back: Coping with the Challenges of Global Electronics. In: Proceedings of the International Symposium on Electronics and the Environment. IEEE: 89 – 92.

Robinson, B.H. (2009). E-waste: An Assessment of Global Production and Environmental Impacts. *Science of the Total Environment* **408:** 183 – 91.

Sthiannopkao, S. and M.H. Wong (2012).Handling E-waste in Developed and Developing Countries: Initiatives, Practices and Consequences. *Science of the Total Environment.* DOI: 10.1016/j.scitotenv.2012.06.088.

Tanskanen, P. (2013). Management and Recycling of Electronic Waste. *Acta Materialia* **61:** 1001 – 11.

Tuncuk, A., V. Stazi, A. Akcil, E.Y. Yazici and H. Deveci (2012). Aqueous Metal Recovery Techniques from E-scrap: Hydrometallurgy in Recycling. *Minerals Engineering***25:** 28 – 37.

United Nations Environment Programme (UNEP) (2009). *Recycling – From E-waste to Resource. Sustainable Innovation and Technology Transfer Industrial Sector Studies.* Available at: http://www.unep.org/PDF/PressReleases/E-Waste_publication_screen_FINALVERSION-sml.pdf. Accessed on 16 November 2012.

Wagner, T.P. (2013). Examining the Concept of Convenient Collection: An Application to Extended Producer Responsibility and Product Stewardship Frameworks. *Waste Management* **33:** 499 – 507.

Widmer, R., H. Oswald, D. Sinha-Khetriwal, M. Schnellmann and H. Boni (2005).Global Perspectives on e-waste.*Environmental Impact Assessment Review* **25:** 436 – 58.

Wath, S.B., A.N. Vaidya, P.S. Dutt and T. Chakrabarti (2010). A Roadmap for Development of Sustainable E-waste Management System in India.*Science of the Total Environment* **409:** 19 – 32.

Zaman, A.U. and S. Lehmann (2011).Challenges and Opportunities in Transforming a City into a "Zero Waste City."*Challenges* **2:** 73 – 93.

PART TWO
MARKET DYNAMICS

3

ENTREPRENEURSHIP AND URBAN WASTES: AFRICAN PERSPECTIVES

M.K.C. Sridhar and T. B. Hammed

Introduction

Wastes in Nigeria, both liquid and solid, have potential recycling values. Municipal solid waste management has attracted considerable interest among communities and within government because of its complex nature and associated economic, environmental and sociological problems (Sridhar and Hammed 2013). These wastes contain a variety of valuable materials that have reuse and recyclable value. As a result, this waste material has afforded many a poor person employment opportunities and a source of livelihood. The materials that have a high reuse value are plastics, paper, metal, textiles, tyres, glass, waste oil and organic matter. Most of the recycling activities are carried out in informal sector, on a small or medium scale. These activities are generally carried out by some particular sections of the population, especially women and children. However, Nigeria has not fully exploited this sector on account of several inadequacies. This chapter identifies how entrepreneurship in recycling can be effectively employed to derive social, economic and environmental benefits, particularly in low-income African countries such as Nigeria.

Global Entrepreneurship in Waste Recyclables

Waste recycling is a global phenomenon arising from resource depletion, unsustainable development, poverty and the current concerns about climate change and its effects. While industrialised countries imbibed the culture of adopting a waste management hierarchy, less developed countries have started looking at recycling as an opportunity for cost savings on products, revenue generation and job creation. Scenarios from some of these countries are given below.

Ghana (Accra)

In Accra, the amount of solid waste generated per day was 750-800 tons in 1994 (Asomani-Boateng 2007), 1,800 tons per day in 2004 and 2,000 tons per day in 2007 (Oteng-Ababio 2010a). Boadi and Knitunen (2003) estimated that residents in low, middle and high income areas generated 0.40, 0.68 and 0.62 kg per day respectively. A study by Asase et al. (2009) on the waste stream in Kumasi, Ghana, showed that organic materials accounted for 63 per cent of waste there. The timber market generates 2,360 tons of sawdust per year. With its moisture content of 36 per cent, it can produce 1,551 tons of briquettes with a moisture content of 10 per cent. The calorific value (CV) is 18.9 MJ/kg (firewood has a CV of 13.1 MJ/kg). The product is sold at 95,000 Cedi per tonne (Anon. 1995). Accra is also known for its waste oil recycling. Some 2,000 litres of waste oil per day go to Korle Lagoon and the market price is $180-200/ton. Odawna Waste Oil produces 2 Mt/day, and the product is used by the power and cement industry as fuel. There is also a booming market in other recyclables and many private entrepreneurs have started small-scale businesses.

Republic of Benin (Cotonou)

According to the 2002 census, Benin's population was 8,500,500, and it has an average ratio of daily waste generated of 0.89 kg per day per person with fermentable organic waste representing the largest proportion of waste materials (45 per cent) (Topano et al. 2011). Waste recycling activities in the country have not been given much attention, probably due to the low population and economic development. However, recently (July 2013) the government established a waste recycling company in Cotonou.The municipality is currently partnered in the project but requires an investment partner to provide the capital needed to complete construction. The country is also known for Songhai, the largest and most productive privately owned farm in the West African sub-region where all organic waste generated is recycled as fertilizer and as an energy source to run the farm.

Box 1: Recycled Bed and Pillow

Bed: Recovered jute from 4½ rice bags (650 CFA), paper strips (from shredded exercise books, which may cost 4,000CFA + transport 1,000CFA) or straw (cheaper than paper) and a craftsman who can sew (labour costs 200 CFA). The empty rice bags are sewn together and stuffed with paper or straw.The frame is made using second-hand beds that have been purchased, repaired and repainted. Bed posts are made from second-hand metal rods; paint and canvas for the bedsprings will cost 3,000CFA; the finished bed will sell for 5,000 CFA.

Pillows are made with new fabric stuffed with paper or cotton waste.A pair of pillows will sell for 700 CFA. Three or four people are employed, and the shed is made of old canvas, with storage space for a few tools, piles of paper and cotton. Other businesses include the repair of folding chairs and the collection of old bottles and any other recyclables.

Senegal (Dakar)

Dakar is noted for its informal recovery networks. The recovery of urban waste occurs primarily outside the modern economic sector, with no official controls. There is currently no formal e-waste recycling infrastructure in Senegal. Plastic waste is partially recycled and ferrous metal and aluminium are exported to advanced countries (Wone et al. 2008). Cardboard boxes, newspapers and non-returnable bottles are collected from watchmen, households, restaurants and hotel employees; rags, flasks and old shoes are also collected from refuse heaps. Some people specialise in empty cement bags and jute rice bags. Approximately 300 people work every day at the dumps using hooks to find usable refuse such as scrap iron, rags, bottles and repairable sandals. They live at the dump sites in houses built from cardboard boxes and sheet metal and work in three different tiers of neighbourhood: wealthy; industrial and middle income; and working class.

There is hierarchy among these people and all these groups collect at "pack" a storage facility where recyclables are transported or sold (Stephanie, D. 1991). "Innovative craftsmen" make a variety of items: packaging materials are made into "gri-gri" a unique fashion amulet; aluminium castings, old cooking pots, motor parts (bolt heads), boxes, metallic capsules and smelted cans are converted into new cooking pots and cutlery. The blacksmith flattens out metal barrels, paint drums, refrigerator sheet metal and car wheel rims to make cooking stoves, pans and other items. Mattress and tapestry makers use a lot of waste as filler (Box 1). New products include briefcases made of Coca-Cola and colourful tomato concentrate cans. Slaughterhouse waste, such as horns, is sold to skilled artisans and carved into lamps and bracelets. Pieces of plastic, rice bags or strips of plastic packaging for shipping are also put to good use. Shoemaking and making bed sheets and children's clothes are other activities involving recycled materials.

Environmental Development Action in the Third World (ENDA): this supports recycling through the creation of jobs associated with recycling activities, and the improvement of the working conditions for informal waste recoveries. In Dakar, there are well organized networks for each major recyclable (ENDA 1991):

(i) ***Metal network:*** In Dakar area, 400 blacksmiths live. Iron and aluminium are the basic materials. Cooking pots (5kg pot 3,000 CFA and 30 kg pot 17,500 CFA), stoves (150 to 500 CFA), skimmers (250 to 300 CFA), kettle mangers (750 CFA), wheel barrows and watering cans are the finished goods. About 54 per cent of these workshops pay a tax of 150 CFA. Complaints from these artisans include insufficient income, a competitive market and unavailability of capital.

(ii) Glass bottle network: There are about 200 waste recoverers, who mostly obtain material from rich neighbourhoods through house boys and maids. The peak season is after Christmas. The suppliers wash the material with soap and hot water before selling, and 25 cl and 75 cl size bottles are in great demand. The average income is about 360 CFA per day. Only 38 per cent pay tax to the government. Many favour the creation of an association. Constraints include inadequate income and unavailability of capital.

(iii) Plastic network: About 50 to 80 waste recoverers were reported. They source the material from hotels and rich neighbourhoods (bottles and drums), traditional neighbourhoods (light sandals), car parks, factories and workshops (tyres and inner tubes). Some of these products are washed with hot soapy water and are in great demand.

(iv) ***Paper/carton network**:* Recovery varies, and animal-drawn carts or head loads are for the main form of transport. Watchmen, maids and utility services are the main suppliers. About 80 per cent of the recoverers pay a daily tax of 150 CFA. The rainy season is particularly bad for business, as the paper is damaged.

(v) ***Wood network**:* About 80 waste recoverers were reported. The main supplier is Dakar Ports Authority and carpenters, and the most wanted products are beams, boards and plywood. No one pays tax.

(vi) ***Main dump site recycling:*** The central dump site receives about 1,000 tons of refuse and industrial waste from 30 industries and provides 300 recycling from. One of the recoveries' remarked: "Around each hut, one finds the refuse of the refuse." This is a reference to the waste remnants left after the recovery process is over. The hierarchy of recycling is metal, paper, plastic, glass and rags. Recyclers dependent on industrial waste are richer than those who depend on domestic recycling. Two entrepreneurs were using BATA leather trimmings

and a retiree from a company (SOADIP) started a scrap metal business. In Baol area, the recycling of rags is predominant, followed by glass, plastics (shoes, bowls and dolls), metal and paper/cardboard. There is still a need to look into the "domestic," "industrial" and "institutional" units in the country. However, it can be said that "the informal waste recovery trade is well organized; it has a firm internal cohesion, though a structural framework is lacking" (ENDA 1991).

Mali (Bamako)

The population of Mali was estimated at 11,419,482 and grows at a rate of 3.7 per cent per year. About 3,551,359 people live in urban areas and 7,868,123 in rural areas (Mamadou et al. 2009). The volume of solid waste produced each day in Bamako increased from 1,500m^3 in 1992 to 2,200 m^3 in 2000, of which only 60 per cent is collected on a daily basis (Modibo 2011).In 2007, the quantity of municipal solid waste secreted in Bamako was estimated at 325,807 m^3 and currently each individual produces approximately 0.65 kg of wastes. Between 2000 and 2002, the city produced an average of 2,000 and 2,200 m^3 of waste per day (Mamadou et al. 2009); only 40 per cent of this amount is collected and each year Bamako produces mountains of municipal solid waste. Most waste remains in the city because of the lack of financial and material means. Refuse in Bamako comprises (per cent dry weight): paper, cardboard (3.5); plastics (2); metals (3.5); glass and ceramics (1); cloth and textiles (12); leaves, straw, wood (17.5); charcoal (2); ashes, sand, dust (51); small quantities of bones and batteries.

Near "Car Cemeteries," traditional craft industries flourish, with people making kitchen utensils, farm tools, toys, etc. The most common form of recycling is sifting household refuse. Sieves with 5.5mm^2 mesh are used to remove sand and dust, leaves and sheep dung. The soil thus separated is enriched with nutrients and has an estimated value of 2000 CFA/m^{3,} excluding transport. This activity is carried out at every refuse dump. The particles remaining after sifting, with low inorganic content, are used for composting. Leftover food waste and oil cakes are used for feeding animals, while groundnut pods and sugarcane trash are used for heating.

Egypt (Cairo)

For a very long time, migrants and squatters in Cairo have systematically sorted through the urban domestic wastes for marketable materials. The bulk of household wastes, such as broken pieces of furniture, piles of unwanted papers, tins, bones, magazines and disused mattresses are not thrown away but sold to itinerant collectors who wander around with push carts (Hekim 1977).

South Africa (Johannesburg)

The 2010 survey aimed at understanding post-consumer recycling behaviour in South Africa does not paint a very positive picture about urban South Africans' attitudes and behaviour when it comes to recycling (Suzan 2013). According to the survey, more than 73 per cent of South Africans living in urban areas reported not recycling at all; about 27 per cent reported some recycling; and only 3.3 per cent of respondents indicated they sort most or all of the selected five recyclables from their household waste and recycle them frequently. The survey results also highlight an overall negative attitude towards recycling. Urban households rarely recycle due to lack of space, lack of time, because it is dirty or untidy, or they lack the knowledge of what is recyclable and what is not. Another reason cited was inconvenient recycling facilities. According to the department of environmental affairs, an estimated 19 million tons of municipal waste was generated in South Africa in 2011, of which about 25 per cent has mainline recyclables. A case study is shown in Box 2.

Recycling in Nigeria

Nigeria, with its estimated population of over 169 million, is noted for solid waste generation and disposal problems. The waste density ranged from 280 to 370 kg/m^3 and the per capita generation of municipal solid waste ranged from 0.44 to 0.66 kg/day (Ogwueleka 2009). Markets, office complexes, motor parks, small and medi-

Box 2: Case Report from South Africa

Recycling Has the Potential to Be Profitable and Create Jobs
(Dimakatso Motau, 2012)

"Currently, recycling does not provide attractive profits ..."and waste pickers are dehumanized by being called "scavengers, pickers," etc. However,

> ... the City of Johannesburg is targeting 4 000 green jobs by 2015 through maintaining and establishing buy-back centres and community recycling initiatives, and formalizing waste pickers. A solid plan needs to be put in place to enforce a structured waste management programme. However, not all waste pickers are keen to be registered, as they want to retain their independence. The municipality only has about 687 registered waste pickers on its database and a pilot project is under way to provide them with protective clothing, such as gloves, glasses and overalls. The second phase of the pilot project is to provide them with sponsored trolleys, as they are particular about the kind of trolleys they want. The City of Johannesburg plans to issue waste pickers with identity cards and get them to form cooperatives. Another aim is to continue hosting knowledge-sharing meetings for the benefit of their growth and our communities.
>
> Materials recycling facilities need to be explored in South Africa but, currently they are expensive to operate, as the country is not on par with its global counterparts in this technology. There is a need for government to accelerate awareness of the potential value of recyclables to improve recycling profitability. This can be achieved through fostering public education and involvement, assisting in establishing outlets to receive reclaimed materials.

um-scale industries generate a variety of wastes. Electronic waste is a new entry. Waste composition has been changing over the years, depending on the economy and government policies. A comparative assessment of the composition of solid wastes in various Nigerian cities is given in Table 3.1. The content of a dustbin is a reflection of the economy and development. In 1960s, after the discovery of oil, the waste was rich in organic matter. The problem was more one of volume of waste due to traditional habits such as food wrapping in leaves, etc.

During the 1980s, plastic replaced much of the metal (copper, brass, etc.) used in households. Aluminium highway dividers have disappeared or been stolen and ended up in recycling industries. By the turn of 1990s, when the Structural Adjustment Programme (SAP) was introduced, economic hardships made people look into their dustbins more critically, and the concept of reuse or recycling began to emerge. A price is put on each waste component in the garbage dump. Itinerary waste pickers and sorters have a field day.

Table 3.1:
Recyclables in wastes from major cities in Nigeria

Recyclables	Nsukka	Lagos	Makurdi	Kano	Onitsha	Ibadan	Maiduguri
Organic waste	56.0	56.0	52.2	43.0	30.7	76.0	25.8
Plastics	8.4	4.0	8.2	4.0	9.2	44.0	18.1
Paper	13.8	14.0	12.3	17.0	23.1	6.6	7.5
Textile	3.1	-	2.5	7.0	6.2	1.4	3.9
Metal	6.8	4.0	7.1	5.0	6.2	2.5	9.1
Glass	2.5	3.0	3.6	2.0	9.2	0.6	4.3

Other	9.4	19.0	14.1	22.0	15.4	8.9	31.3
Total	**100**	**100**	**100**	**100**	**100**	**100**	**100**

Source: Ogwueleka (2009)

Currently, besides the organic component (60 to 80 per cent), plastics (15 per cent) and metal (10 per cent) dominate in waste. In some locations, for example where oil and industrial multinationals operate, the lifestyle of workers has changed significantly, and as a result recyclables emanating from packaged foods and beverages have exploded. These areas are in Port Harcourt, Bonny Island, Yenagoa, to name but a few. There have been occasional studies, but no comprehensive study for a long time. Ayorinde et al. (2010) indicated the mean generation rate in Ibadan to be 0.452 (low income), 0.303 (medium income), 0.688 (high income) and 0.636 (rural) communities. Box 3 shows the predominant recyclables in the waste.

Recycle and Reuse Markets

In many urban centres, markets exist that deal in goods that have been recycled or are meant for reuse. A quick appraisal was carried out in Ibadan (Sridhar et al. 1997) involving 14 traders dealing in such materials such as: clothes, rags, paper, metal containers, scrap, tin, bottles, glass, shoes, rubber or plastic slippers, nylon bags, spring beds, foam mattresses, and a host of others. They were mostly men,64.3 per cent of them Muslim and 35.7 per cent Christian, and had one to 10 years of experience in the trade. Most

Box 3: Recyclables in Ibadan (Ayorinde et al., 2010)

Bottles	82%
Plastics	46.5%
Nylon	21.1%
Shoes	36.9%
Rubber	21.9%
Metal	31.6%
Textiles	9.6%
Batteries	7.9%

were youths, middle aged and were married. The mother tongue was Yoruba (85.7 per cent). They had formal education at primary or secondary level and most can communicate in English in addition to their native tongue. They had associations, the main being the Scraps Association of Oyo State.

Most of the metal scrap materials were procured from mechanical workshops (21.4 per cent, mostly scrap), collected from recycling markets (14.3 per cent, slippers, plastics, tins, bottles, mattresses and springs), Badagry, Yaba and Aba markets (21.4 per cent, clothes, shoes and belts), industries (14.3 per cent, bags for sugar, cement, flour, salt, guinea feeds, glass), office premises (14.3 per cent, paper, etc.) and dumping sites (14.3 per cent, nylon, plastic, polythene bags, tins, bottles, slippers, lanterns). These materials were either bought or procured by dealers directly (57.1 per cent) or through agents, relatives, employees or others (42.9 per cent). The quality of these materials is important for their resale value. About one-third of the dealers were satisfied with the quality and rated the materials as fairly good and about 21.4 per cent of them rated them as good. Iron and aluminium scrap was sent to Saki, where there are well organised recycling industries, which make cooking pots, cutlery and other domestic utensils. Glass was sent to Toyo Glass at Agbara industrial estate in Ogun State. Plastics were recycled in many industries in Ibadan, Lagos, Sango-Ota and elsewhere. Regarding economic value and profitability, 85.7 per cent of dealers said they earned "barely sufficient to survive on," which seems to be a conservative statement.

Scrap-Collection Practices for Recycling

A study carried out by Sridhar et al. (1992) indicated that the people who collected recyclables from dump sites fell into the 9-15 year age group, were generally male and had a primary or lower secondary education. Those who collected from households were generally older women. The collection of recyclables occurs at three levels – primary, secondary and tertiary. At the primary level, recyclables are collected by numerous individuals for the purpose of reuse or sale for conversion into other useful articles. Thus, plastic containers and bottles are used to carry water or kerosene; margarine or milk tins to make traditional foods such as *moin-moin* (cowpea/bean cake). From the primary level, the recyclables are purchased by adults. They wash or clean the materials, pack and transfer them to the final destinations, which are small-scale industries. At the tertiary level, these small industries further process the recyclables and convert them into goods. In market and residential areas, old women and the destitute can be found collecting left-over beans, maize and other food items for immediate consumption. Many of these waste pickers are too shy to talk to other people. They account for a large fraction of the informal sector in waste management, but have yet to be recognised. More recent surveys were carried out by Sridhar (2012) in Yenagoa (Bayelsa State) and by Hammed and Abegunrin (2012) at four Ibadan dump sites between 25 May and 8 June 2012. The results indicated that recycling activities were very lucrative at the dump sites. Batches of recyclables, comprising about a dozen components, were systematically segregated and carried to nearby cities

Figure 3.1:
Private-sector collection of recyclables at a refuse dump
(Source: the authors)

for sale to middlemen, who in turn sell to industry (Figure 3.1).

Recycling Options for Nigeria

In Nigeria today a large number of industries have started to process raw materials derived from recycled wastes. The range of industries that have sprung up is evident from the Industry Directory. These industries incorporate recycling as part of their cost-reduction processes (Figure 3.2). Paper is no longer wasted while scraps and PET bottles have disappeared from waste. These items are taken care of by several small and medium industries in major cities. Bones are recycled or exported, while batteries are recycled for their lead and plastic content. The transport sector also benefited from these activities.

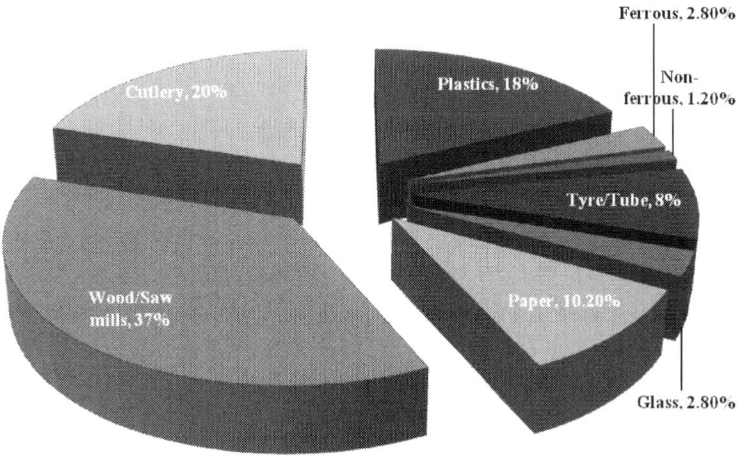

Figure 3.2:
Nigerian industries that incorporate recycling into their processes

Source Separation of Recyclables

There has been growing concern about source separation of recyclables, which is widely believed to be a cost-effective approach. In high-income countries, the motivation for source separation is to be found in the scarcity of landfill space and the financial and environmental costs of disposal. In low-income countries, there is a demand for separated materials in line with traditional practices of selling, bartering or giving unwanted resources. There has been no comprehensive information on such activities among various income groups at the global level, only sporadic reports. In Nigeria, Sridhar and his group carried out hands-on training on source separation of recyclables in six communities in Ibadan and two in Lagos. The exercise took place over 30 days and the various communities were mobilised. However, this activity could not be sustained, as there was no support from state or local governments. Currently, Lagos is trying to introduce kerbside kiosks to separate recyclables as a pilot projects. Multinational oil companies such as Shell and Chevron, however, are practising source separation in their workers' camps.

Source separation is ideal for material recovery, particularly when it is carried out at the point of generation and before items get soiled. India has been promoting source segregation as a means of zero waste management. In Quetzaltenango, Guatemala, waste oil is being collected from automobile mechanics and reused. In Sweden, wet/dry collection has been promoted, whereby dry waste goes for incineration and wet waste is used for composting. In Sri Lanka, the National Forum of Peoples Organisations (NFPO) operates a compost site that recycles the entire fruit and vegetable waste from a market in a town with a population of 20,000. Here, traders separate their organics prior to disposal. In many provinces in Argentina, the organic part of wastes is collected in 5 kg quantities, taken to a designated platform and composted using earthworms (*Eisenia foetida*). In 1992, only 400 families were involved in this project, but by 1997 the number had risen to 22,000.

Table 3.2 shows the differences between source separation and the centralised separation of municipal solid waste. A large private glass company in Sri Lanka established a waste glass collection mechanism involving the participation of the urban poor. It relied on issuing loans to the poor at favourable interest rates to purchase cycle-trailers, which are then used to collect glass from households. Cities and towns in developing countries cannot be expected to view waste reduction with the same urgency as those in more wasteful economies, and often are much more preoccupied with the basic, efficient operation of collection and disposal.

Among the main factors affecting waste generation and recovery/reuse/recycling are the: standard of living and lifestyle of consumers; retail prices; materials scarcity or costs for industry; diversity of local and regional industry; existence of social groups associated with waste work; numbers of rural migrants seeking work and numbers of street people; capacity to trade materials beyond the local economy; existence of intensive farming in peri-urban areas; availability/cost of fertilisers, animal feed and organic matter; technology and efficiency of official waste collection and disposal systems; politics of local authorities regarding waste workers; and public awareness of waste management procedures and issues.

Table 3.2:
Differences between source separation and centralised separation of municipal solid waste

Source Separation	*Centralised Separation*
Better product quality	More contaminated products
Increased waste awareness	Shielding waste awareness
Complex logistics	Easier logistics
Public involvement needed	Less involvement needed

Source: Reichenbach (2005)

Organic Recycling and Organic Agriculture

With the recent announcement by the Nigerian federal government banning imported fertilisers, there is a tremendous impetus to recycle organic wastes from residential areas, market places and industries. Though there are no large-scale composting plants operating in Nigeria today, there have been attempts to promote indigenous technology. A simple semi-mechanised system of windrow type is ideal for the country. The inputs are raw sorted waste; labour; minimal hand tools such as spades, wheel barrows, water sprinklers, sieves, and a tipper to carry the raw materials and finished product; and mechanical components such as a sorter, pulveriser and mixer. It is recommended that such plants be encouraged at community level, local government level and within institutions. As experience has shown in Nigeria and several other countries, highly sophisticated imported units are unlikely to work effectively.

Human resource management is a major problem, and, as much as possible, community or NGO involvement should be encouraged. Existing waste disposal agencies may be persuaded to adopt composting as one of their disposal strategies on a trial basis, and this system can be perfected. Environmental education and enlightenment are required to promote this technology. In our experience, farmers' groups in Kwara, Ogun, Oyo and many northern Nigerian states have tried our products and have been disseminating the technology among colleagues in their states. Composting is indigenous to Nigeria, but has been neglected. It is high time that this old and appropriate technology be revived.

Technologies for Recycling Entrepreneurship – Indigenous versus Imported

Among the many factors threatening the success and sustainability of waste recycling ventures in Nigeria, technical and mechanical factors appear the most conspicuous. Other factors are financial, institutional, political, legal, physical/operational, economic and psychosocial constraints. Many a time, recycling machines are imported into the country and hardly work effectively for a year. Many officials in charge of solid waste management, particularly at local level, have little or no technical background or training in engineering, management or public health. Similarly, research and development activities in solid waste management are often a low priority in developing countries, leading to the selection of inappropriate technology in terms of local climatic and environmental conditions.

In one case, nylon recycling machines were imported by the government and located at landfills across the country (Figures 3.3 and 3.4). Until now, the machines have not been put into use because of technical constraints and other factors arising from the ensuing political imbroglio. At Alesinloye market in Ibadan, the authors have constructed and installed similar machines, using locally available technology and materials. These machines have been in operation for more than six years. In addition, imported machines have very expensive spare parts, which may affect their choice by and preference among waste recyclers in developing countries. In these countries, wastes are just mixed and segregation into different components is not yet routine practice. Importation of machines and technology connotes over-dependence on foreign experts. Meanwhile, there are sustainable and dependable technologies that can cope with the unique nature of the waste and the variations in weather conditions. These indigenous technologies are very simple and can be promoted to all tiers of government. They are not too sophisticated or automated, and have great potential in developing people's capacities and generating wealth. Using these technologies, the authors have undertaken various community-based projects across the country. These include: plastics recycling into plastic chips and ebonite rubber; nylon recycling into nylon pellets and flakes; farm residues into smokeless charcoal; sawdust into wood composite; organic waste into organic fertiliser and biogas for cooking and electricity; and scrap metal into ingots and tools.

Figure 3.3: A locally fabricated nylon recycling machine at Alesinloye market, Ibadan (left) and **Figure 3.4** A brand new imported machine dumped at a landfill site at Sango Ota, Ogun State (right).

***Source*:** Authors' unpublished work.

Key Actors and Activities in Waste Recycling Entrepreneurship

Recycling and/or reuse activities are known for their economic, environmental, health and social significance. The key issues involve policy, reduction of volume of waste, maximising separation at source, promoting small-scale waste recycling industries and resource recovery and formal waste management. In implementing these issues, the key actors need to be identified. Public awareness campaigns, mobilising community participation, providing incentives and training and promotion are the key inputs for a waste recycling venture to achieve its primary objectives, including making a surplus and environmental sanitation. The principal actors are:

- *Domestic Consumers and Industry:* They influence the amount and the nature of wastes.
- *Government:* Sets priorities through regulations and laws, in relation to consumers and industry.

To obtain maximum profit in a value chain, various levels of intervention are required to address the following:

(i) **Creation of awareness:** Awareness of environmental and public health issues will promote structural improvement and environment-friendly behaviour by people and encourage more community participation in recycling and reuse. Some effective methods are use of traditional and modern community communication channels; targeting youth groups, which are prone to be attracted to changing lifestyles; mass media; and by organising workshops and meetings in neighbourhoods. These campaigns should take place at the beginning of a scheme and throughout the project period. The components should include education and information and local politicians and policymakers should also be involved. Government, NGOs and educational institutions may be the appropriate facilitators.

(ii) **Incentives:** These may be used to obtain positive results in the short term. Since these have an economic character and government may have to spend money, it is important that results be seen in a short period.

(iii) **Restrictive legislation:** When awareness and incentives fail, legislation is the option. This also demands monitoring and sanctions, and will have substantial costs for government.

(iv) **Community mobilisation:** This will promote waste separation and effective collection. It can achieve short- and medium-term results. Workshops, neighbourhood communication sessions and financial incentives will encourage better participation. "Who should participate? What is expected of them? To what extent they should contribute?" These are some of the questions to be kept in mind in formulating this strategy. Support should be obtained from the formal waste management authority and people should be involved right from the beginning. One of the important outcomes is establishment of neighbourhood associations. The general education level should be high for separation of wastes to be achieved and community participation can only be secured through the creation of understanding and appreciation. Also, although community participation should be initiated at neighbourhood levels, it should be coordinated under the auspices of local authorities. In low-income areas, separation of wastes at household level is very profitable. Door-to-door collection of useable materials can be arranged, or, alternatively, households can bring their collected paper, glass and plastic bottles to transfer depots or collection points, where dealers could have their shops or market stands.

(v) **Training and capacity building:** Transfer of know-how should be an important component of government programmes. The training should include recycling techniques, organisation, marketing, communication and health aspects. There is a general lack of information for those who operate or initiate small-scale waste recycling industries and therefore there is a need to provide more. The success of recycling depends on the marketability of products, as people have a preference for products from virgin materials. Training and promotion should lead to action plans. The required inputs for the success of training activities are establishment of pilot projects, long- and short-term assistance and financial support.

Communities in Recycling Entrepreneurship

The role of waste recovery in livelihood creation demands the attention of local governments, non-governmental and community-based organisations in developing world cities (Christian 2001). Recycling and reuse cannot be emphasised without waste minimisation as a prerequisite. The economic assessment of recycling and reuse is difficult because some of the economic benefits are hard to measure. Indirect economic benefits such as health and environmental costs are also difficult to quantify. Recycling solid wastes is labour-intensive. Urban centres in developing countries have relatively numerous migrants and unskilled workforces, which may be tapped effectively. This will at once generate employment and in turn reduce poverty. However, the quality of the urban environment is directly related to the general public health status of the residents. People involved with waste recycling and reuse are a high-risk group. They often suffer from diseases such as eye irritation, respiratory illnesses and skin disease and are exposed to smoke, dust, airborne pathogens and parasites, to name just a few health hazards.

Many small-scale industries are dependent on recyclers and waste pickers for their supplies of cheap raw materials. The starting point for understanding the social dimensions of waste recovery is recognition of the centrality of waste recycling in the lives of the poor. Different groups are involved in waste recycling: householders, door-to-door collectors, formal waste collectors, street-waste pickers and dealers and traders in wastes. Waste pickers are the poorest, and are thus highly dependent on the intermediaries who exploit them. They are the victims of discrimination, and are often branded as thieves and delinquents. A study conducted by the UN's Human Settlements Programme (UNCHS) revealed the following income distribution among participants: waste pickers, US$ 44; street peddlers, US$ 60; small-scale, cottage-based waste-recycling industry, US$ 100; and dealers, US$ 385. This gap is too wide. Beyond that, it is estimated that about 1 per cent of urban dwellers make a living from recycling. Upgrading the recycling sector will double employment and improve working conditions.

Community participation and mobilization should therefore address some of the key issues below:

(a) **Waste minimisation:** This can be achieved by changing consumption patterns and lifestyles, and using more recyclable materials while banning waste imports.
(b) **Maximisation of source separation:** This can lead to employment creation and better working conditions. It also leads to availability of more uncontaminated waste materials and increased waste management efficiency.
(c) **Promotion of small-scale waste recycling industries:** This increases the level of knowledge and the development of local technologies.
(d) **Integration of recycling and formal waste management:** This can be achieved by improving communication methods, increasing management capabilities and developing monitoring instruments.

Conclusion and Recommendation

Waste recycling and reuse have become a lucrative informal-sector venture across the globe. To make a really successful venture of waste, all the stakeholders – government, host community members, machine fabricators, researchers, waste recyclers, consultants, environmental health professionals and others have great roles to play. The approach should be holistic to be sustainable. It should include community education; mobilization and awareness through the media, seminars and workshops; enforceable government regulations and legislation; training and retraining of supervisors and workers at plants; and use of incentives. Also, entrepreneurship opportunities in waste should be explored in the context of local initiatives and prevailing local environment to reduce over-dependence on foreign technology that may not be sustainable in many African countries. The four points outlined in the previous paragraph also constitute our recommendations. To them we would add the importance of Public-Private Partnerships (PPP) in promoting successful entrepreneurship.

References

Anomanyo, E.D. (2004). Integration of Municipal Solid Waste Management in Accra (Ghana): Bioreactor Treatment Technology as an Integral Part of the Management Process. Dissertation, Lund University. Available at www.lumes.lu.se/database/alumni/03.04/theses/anomanyo_edward.pdf>.

Asase M., E.K. Yanful, M. Mensah, J. Sanford and S. Amponsah (2009). Comparison of Municipal Solid Waste Management Systems in Canada and Ghana: A Case Study of the Cities of London, Ontario and Kumasi, Ghana. *Waste Management* **29**(2): 779 – 86.

Asomani-Boateng, R. (2007). Closing the Loop: Community-Based Organic Solid Waste Recycling, Urban Gardening, and Land Use Planning in Ghana, West Africa. *Journal of Planning Education and Research* **27**(2): 132 – 45.

Ayorinde, D., M.K.C. Sridhar, W. Bolanle, S.O. Asamu and T. Tajudeen (2010). A Report on Characterization of Solid Wastes in Ibadan: A Rapid Appraisal Survey in Four Residential Communities Conducted Between August 4 and October 24, 2009. Oyo State Ministry of Environment and Water Resources/InWEnt Capacity Building International Waste Management Programme, Ibadan.pp. 1– 60.

Boadi K.O. and M. Kuitunen (2003).Municipal Solid Waste Management in the Accra Metropolitan Area, Ghana. *The Environmentalist* **23**(3): 211 – 18.

Christian, M.R (2001). The waste sector and informal entrepreneurship in developing world cities. *Urban Forum* **12** (2): 247 – 59.

Dimakatso, M. (2012).Recycling has the potential to be profitable and create jobs. Waste Management 1, Creamer Media's Engineering News Online, http://www.engineeringnews.co.za. Accessed 30 November 2013.

El-Hekim, S.M. (1977). The role of the family kinship and rural/urban migration in the processing of solid wastes in Cairo, *Africa Development* **11**(2): 115 – 22.

ENDA (Environmental Development Action in the Third World) (1991).*Man and Waste: Popular recycling activities in the third world*, M.M. de Botero (ed). ENDA: Dakar. pp. 1 – 297.

Hammed, T.B. and J.A Abegunrin (2012).Waste characterisation and GIS mapping of recyclables at designated dumpsites in Ibadan, Nigeria.MPH student project, Department of Environmental Health Sciences, University of Ibadan.

Mamadou, S. T. Zhonghua, H. Win and W. Jun (2009). State and Management of Solid Wastes in Mali: Case Study of Bamako. *Environmental Research Journal* **3**(3): 81 – 6.

Modibo, K. (2001). Building partnerships for urban waste management in Bamako.http://pubs.iied.org/pdfs/9123IIED.pdf. Accessed on 14 December 2013.

Ogwueleka, T.C., (2003). Analysis of urban solid waste in Nsukka, Nigeria. *Journal of Solid Waste Technology and Management* **29**(4): 239 – 46.

Oteng-Ababio, M. (2010).Private Sector Involvement in Solid Waste Management in the Greater Accra Metropolitan Area. *Waste Management and Research* **28**(4): 322 – 29.

Sridhar, M.K.C., A.A. Ajayi and A.M. Arinola (1992). Collecting recyclables in Nigeria, *Biocycle* **33**(7): 46 – 47.

Sridhar, M.K.C. (1996). Turning wastes to wealth: A strategy for urban poverty alleviation in Nigeria. In *Proceedings of a Workshop on Urban Waste Management: Appropriate and Affordable Technology and Strategies*, A.G. Onibokun (ed). Centre for African Settlement Studies and Development (CASSAD), July. pp. 86 – 96.

Sridhar, M.K.C. (2011). Developing Countries Must Look Inwards to Move Outward: An Analysis on Efficient Resource Utilisation. Proceedings of the World Resources Forum 2011, EMPA Materials Science and Technology, St. Gallen. Organised at Davos, Switzerland. pp. 1–6

Sridhar, M.K.C. and T.B. Hammed (2013). Turning Waste to Wealth in Nigeria: An Overview. *Journal of Human Ecology* **46**(2): 195-203.

Stephanie, D. (1991). *Valorising urban waste to create employment, in man and waste: Popular recycling activities in the third world*. Dakar: ENDA. pp.101 – 110.

Suzan, O. (2013). South Africa's household recycling target. Source: Council for Scientific and Industrial Research. Available at http://www.southafrica.info/about/sustainable/waste-070113.htm#.Uq33fM7N3yU#ixzz2nZQq9QAe.Accessed on 12 December 2013.

Topano, N., M. Domeizel, J. Fatombi, R. Josse and T. Aminou (2011).Characterization of Household Solid Waste in the Town of Abomey-Calavi in Benin. *Journal of Environmental Protection* **2**(6): 692 – 9.

Wone, S.S., D. Rochat, C. Gassama and C. Kane (2008). Senegal: E-Waste Country Assessment. *Proceedings of the 19th Waste Management Conference of the IWMSA (WasteCon2008), 6-10 October*. Durban. ISBN Number: 978-0-620-40434-1.

4

DREAMING OF A "ZERO WASTE CITY" INSIGHTS INTO "KAYA BOLA" AND THE STEEL INDUSTRY IN ACCRA, GHANA

Martin Oteng-Ababio

Overview

Global waste is changing dramatically in volume and content, a process fuelled by the increasing world population and globalisation. In Accra, the authorities see the littered streets and filthy settlements as an impediment to public health. Yet through investment in economic development and the city's envisioned transformation into a "millennium" city, Accra is said to be in unique position to create a new standard for cities on the continent. Thus the work of "Kaya bola," those people in the informal sector, who subsist by removing and recycling "waste" materials for resale or reuse, assumes greater environmental and economic importance. This paper explores the role and working conditions in the sector, and how participants reconnect with the formal steel industry. The findings show that these activities provide participants with increased income, the steel industry with cheaper raw materials and the city with cheaper but efficient environmental services, albeit under compromised working conditions including social stigmatisation and incessant harassment.

Introduction

Generally, urbanisation and urban growth are intricately intertwined with economic growth, but the future of cities depends on how the rate of urbanisation, and the accompanying increase in waste generation, is handled, particularly in Asia and sub-Saharan Africa, where most extreme poverty is concentrated (Songsore 2010; McGranahan and Martine 2012). Though cities cover only 2 per cent of the earth's surface, they generate 70 per cent of world's waste. With the world population expected to increase by one billion by 2030, cities will have an increasing role to play in creating "liveability" (UN Habitat 2010). Ineffective and inefficient waste management, besides its negative impacts on health (through increased air, land and water pollution), results in greenhouse gas and toxic emissions and thc loss of precious materials and resources.

Most developed countries seem to have surmounted this hurdle, rediscovering waste from a resource management perspective, with few having the luxury to advocate for a "zero waste" city. "Zero waste" means managing products and processes systematically to avoid and eliminate waste and materials, and to conserve and recover all resources from waste streams (Zaman and Lehmann 2011). Achieving zero waste by 2030 is an expressed goal of Los Angeles (see http://www.zerowaste.lacity.org) but from a holistic point of view designing zero waste cities is relatively hard to do. In developing countries where solid waste management has traditionally been a major challenge, notwithstanding notable local government interventions (Fahmi 2005; Okot-Okumu and Nyenje 2011; Nzeadibe 2009), the increasing urbanisation of poverty and decreasing availability of and increasing prices for raw materials have made the "resources" contained in waste more economically interesting.

In most urban landscapes, e-scrap collectors, for example, in spite of their marginalised position, increasingly see waste as an important economic resource to which they seek to gain access (Parizeau 2011). Their ability to recover recyclable and reusable materials, amounting to up to one-third of the waste destined for dumping or dumped, and to do so in a self-financing way using rudimentary methods, allows them to mitigate their poverty, provides valuable feedstock to industry and reduces the 2,200 tonnes of waste produced daily in Accra (Oteng-Ababio 2010; 2011a; 2012). Their immense contributions to the urban economy and efficient solid waste management notwithstanding, they remain unrecognised and operate in a precarious, low-paid, highly stigmatised environment, pregnant with health risks (Parizeau 2011).

This chapter contextualises these issues based on lessons learnt from the Kaya bola in Accra, which is regarded as an exemplar among the fastest-growing cities in Africa (Grant 2009; Oteng-Ababio 2011b). It explores how insights gleaned from the relationship between the Kaya bola and the steel industry can illuminate our understanding of the relevant issues theoretically and empirically and possibly help to transform traditional solid waste management practices towards the attainment of a zero waste city. Going beyond an emotional presentation of the problem, the chapter attempts to understand the complex drivers of poor solid waste management practices, instead of propounding a "one size fits all" solution (Obeng-Odoom 2011).

The chapter forms part of broader research into the e-scrap circuitry of the urban economy of Accra, based on a survey of 89 Kaya bolas and interviews with 22 key stakeholders carried out in 2011. The author has since returned to conduct further interviews to fill in the gaps and interrogate emerging issues. Such interviews were designed to target workers who collect scraps primarily from waste dumped or destined to be dumped, as opposed to collecting exclusively from larger institutions, such as the Ghana Police Motor Transport and Traffic Unit (MTTU), mines and industry. Experience of years of research in the city helped in determining the areas frequented by scrap collectors. During data collection, survey participants were randomly selected, and because the research team included a well-respected community leader, the refusal rate was almost zero. Interviews with key informants, who included public officials and community leaders, examined the prospects and the challenges facing the e-scrap industry.

Below, I first discuss the concept of a zero waste city as popularised in 2004 by the working group of the Zero Waste International Alliance (see http://www.zerowaste.lacity.org). This is followed by analysis of the policies and key players in the e-scrap industry, highlighting the characteristics, challenges and adaptive strategies, before assessing the e-scrap circuitry. The spatial contours of the broader e-scrap operations are discussed and issues raised as to how "waste" materials are transformed and how elucidating such issues can inform the search for the zero waste city.

Interrogating the zero waste concept

Most developing countries are undergoing unprecedented urbanisation. Cities have at one and the same time become centres of innovation, democracy and economic activity *and* of over-consumption, increasing their per capita waste generation compared to low-consuming cities (World Bank undated). Inadequate urban management, often based on inaccurate information, can turn opportunity into disaster, hence the quest for the zero waste city. Conceptually, zero waste, which is an extension of the 3Rs (reduce, reuse and recycle), is a philosophy and a goal. As a philosophy, it encompasses solid waste management practices that aim to reduce the amount of waste to zero by redesigning resource-use systems. It contests the linear waste management doctrine that creates the need to throw away things, and strives to extend current recycling and reuse practices into a circular waste system.

In principle, the concept "maximizes recycling, minimizes waste, reduces consumption and ensures that products are made to be reused, repaired or recycled back into nature or the market place" (Wes Muir 2000: 1). The ultimate objective of zero waste is fundamental change (not an end-of-pipe solution), where products are designed so that their components can be dismantled, repaired and/or recycled. A working definition adopted by Zero Waste International Alliance in 2004 encapsulates all the tenets and goals of the concept.

> Zero Waste is a goal that is ethical, economical, efficient and visionary, to guide people in changing their lifestyles and practices to emulate sustainable natural cycles, where all discarded materials are designed to become resources for others to use. Zero Waste means

> designing and managing products and processes to systematically avoid and eliminate the volume and toxicity of waste and materials, conserve and recover all resources, and not burn or bury them. Implementing Zero Waste will eliminate all discharges to land, water or air that are a threat to planetary, human, animal or plant health. (http//www.zerowaste.lacity.org)

The concept ends the idea of managing waste, introduces the concept of managing resources and strives to eliminate waste. It connotes an end to subsidies for wasting and making manufacturers assume responsibility for the entire lifecycle of their products. Recent literature from the global North and emerging countries is replete with cities being designed on "eco-city" concepts: Vauban Freiberg (Germany), Hammarby Sjöstad (Sweden), Masdar City (UAE) and Tianjin Eco-City (China) (Lehmann 2010a: 111).

The global applicability of the concept is beyond the scope of this chapter, and a few comments will suffice. Most current eco-cities have population densities of between 50 and 150 people per hectare and it may not be possible to accommodate the entire world's population in such a limited global land area using the same design criteria. Additionally, a true eco-city should not have a large ecological footprint. In reality, however, most cities today are not only over-populated and over-consuming, but they are also depleting finite global natural resources at a high rate (UN Habitat 2008: 24). Thus, to say that managing expanding cities in a sustainable manner is crucial to ensuring global sustainability is an understatement (Vezzoli and Manzini 2008: XI).

As already noted, Accra produces approximately 2,200 tonnes of waste daily (Oteng-Ababio 2010; Owusu et al. 2012), with each of its almost three million residents and a floating population of over one million generating approximately 0.65 kg of waste per day (Oteng-Ababio 2011b). Prior studies have revealed that the authorities, in their pursuit of a clean, liveable city (Oteng-Ababio et al. 2012), have implemented two main arrangements for solid waste management – the house-to-house (HH) system in affluent areas and communal container collections (CCC), expected to be emptied twice daily, in low-income and commercial districts (Oteng-Ababio et al. 2013). These arrangements notwithstanding, waste management is an increasing challenge, a situation UN Habitat (2010) attributes to institutional, technical and financial constraints.

The failure of public sector delivery has created an economic niche for the removal and recycling activities of the Kaya bola (Obirih-Opareh and Post 2002; Medina 2005). Historically, these activities have been tolerated or have gone without comment for two main reasons: first, the numbers of participants were limited, and second, their activities took place mainly at dump sites out of public sight (Adama 2012). In recent years, however, the persistence and growth of informal recycling, in tandem with a long trajectory of increasing poverty and unemployment, particularly after the introduction of Structural Adjustment Programme (SAP) in 1983, has been phenomenal. By way of context, formal sector employment reduced from 484,000 in 1985 to 186,300 in 1991, representing an annual average decline of about 3.7 per cent, with the Cocoa Marketing Board alone retrenching 40, 000 workers in 1985 (Aryeetey and Tarp 2000).

The mass expulsion of immigrants, including over one million Ghanaians, from Nigeria in 1983, which reflected "pendulum-like shifts in the relative fortunes of Ghana and Nigeria as poles of attraction for migrants" (Van Hear 1998: 78), worsened the situation. The sustained economic decline and financial crisis at the time was characterised by a decisive shift in informal recycling work. With inflation running at 100 per cent annually, imports of raw materials for industrial use became prohibitively expensive (Aryeetey and Tarp 2000). This situation created an increased demand for recycled materials from local markets, and also gave rise to soaring unemployment. Joining the ranks of the Kaya bola was thus personal response to labour insecurity and these people became a source of affordable raw materials for local steel industries.

With e-scrap collection becoming increasingly visible not only at dump sites but within the city, Kaya bola came under an umbrella, the Greater Accra Scrap Dealers Association, which, according to Grant and Oteng-Ababio (2012), offered them a sense of social legitimacy, stability and economic value. It also enabled them to scale-up the volume of collected materials and served as a collective basis for interactions with city authorities, which had persistently threatened to evict the Kaya bola from their places of work. It further offered a platform for activism on pertinent policy decisions and for creating a cohesive force to successfully lobby against the twin scourges of forced eviction and deteriorating living conditions.

By 2010, e-scrap collection had become highly visible and, according to the association's chairman, is the mainstay of the 7,000 registered members and also many urban and peri-urban residents. Their contribution and indispensability to both the solid waste management system and local industrial circuits was beyond dispute (Grant and Oteng-Ababio 2012; Amankwah 2013). Regardless, their relationship with city authorities and some residents continues to be marred by confrontation instead of collaboration (Oteng-Ababio et al. 2013). This ambivalence stems in part from the increasing allure of foreign approaches to solid waste management to local officials.

Recent discourse on city waste management seems to have responded to this trend by calling into question the efficacy and appropriateness of the wholesale adoption of foreign-based technologies that are often incompatible with local physical, social and economic realities (UN Habitat 2010: 4). The call has economic legitimacy, as there are few success stories but enough lessons on the pitfalls and problems such inappropriate policies can bring. Meanwhile, the Kaya bola experience offers an inspiring example of how to seize "home grown" opportunities in dreaming of a zero waste city (UN Habitat 2010: 24). Thus, the integration of the Kaya bolainto urban governance structures can improve waste diversion rates and hasten progress towards a zero waste city and/or create more sustainable solid waste management systems, though not without difficulty.

Data and Discussion

Agbogbloshie: Circulating, recirculating and transforming materials

Agbogbloshie is a general scrap-processing zone that houses a wide range of workers, including those dealing in scrap metals (e.g., car parts, old batteries). The settlement (also known as Sodom and Gomorrah) appears to be a shining beacon in the pursuit of the zero waste city, as it has transformed its image from that of veritable hotbed of crime into a flagship of grassroots power and participation. This study builds on earlier ones (Oteng-Ababio 2010; Grant and Oteng-Ababio 2012) and demonstrates how e-scrap does not necessarily follow the path of waste, recycling, pollution and final disposal (Coe et al. 2008). Rather, it exhibits multiple "wasting" and "valuing" unconformities with level of technology (know-how) and geographical location.

The study collected official trade data on exports of e-scrap – mixed scraps, lead, copper –through the Ghana customs, excise and prevention services data management system (GcNET/GCMS). This system compiles and aggregates trade data from all official entry points, capturing information on the origin, importer and exporter (consignee) and type of commodity, among other details. Also as part of the study, some of the industries operating in the field were interviewed, but gaining access to "foreign" companies, especially those in the free-zone enclave, required an introduction by a senior customs official. This is because these companies felt that their previous exclusivity agreement with the government has been abrogated as a result of trade espionage by former employees. In addition, 27 in-depth interviews with key stakeholders in the industry – customs officers, public officials, Scrap Dealers Association executives, collectors and middlemen were conducted. Both secondary and interview data were compiled into themes regarding how much e-scrap is re-circulated and the possible impact of that on the quest for a zero waste city.

Local policy context and myriad players

Trade in e-scrap is subject to an array of local and international export restrictions, ranging from taxes to outright bans on certain exports (e.g., ferrous metals). The official justification for the ban on exporting ferrous metals is that it is to ensure the domestic economy has adequate supplies (i.e., resource conservation), while also protecting the environment and abiding by international trade standards. Table 4.1 examines the sector by location, nature of business, mode of operation and coping strategies.

Free zone companies in Ghana enjoyed exclusivity in e-scrap exports from 2004 to 2010, due to a confluence of state policies that virtually created a "state-sponsored monopoly" for these foreign-dominated firms. The Ghana Free Zone (EPZ) Act, Act 504 of 1995, shifted the emphasis from central planning towards market forces and trade liberalisation, and provided incentives to manufacturing firms. The dearth of knowledge about the real value of "waste scraps" also aided profit-motivated foreign firms in lobbying the government for exclusivity in key "grey areas." In 2004, the government acceded to this demand, and the companies operated in "secrecy," which they maintained by employing expatriate management staff. However, their dominance was soon challenged by certain local actors, hitherto virtually gagged by the government's exclusivity "clause."

Indeed, this challenge gave birth to the illegal exportation of scraps obtained by grinding motherboards into powder as a way of outwitting the "law" and customs officials. This illegality ran in parallel with intense protestations by local actors with insight into the economic potential of the e-scrap business against the agitation by free zone companies about breaches of the free zone Act. To mitigate the problem of e-scrap smuggling, government enacted the Exportation of Non-Ferrous Scrap Metal Regulations (LI 1969) on 15 June 2010. This revoked the earlier monopoly and granted local companies the opportunity to export scrap metals.

Table 4.1: Characteristics, challenges and adaptive strategies of e-waste recycling firms in Ghana

	Characteristics	Challenges	Coping strategies
Free-zone firms	Registered with the registrar general's department and Ghana investment promotion centre Foreign dominated (subsidiaries of parent companies) Capital intensive Enjoyed exclusivity in the e-scrap trade until 2010 Work normally through agents Enjoy "easy" access to capital/technology (home/abroad) Collude as much as possible	Stiff competition from local enterprises Relatively high overhead costs High investment in land, building and equipment	Engage agents within the West African Sub-region (Burkina Faso, Niger, Côte d'Ivoire, etc.) Diversify product line (lead, iron rods) Lobby public officials and sponsor public advocacy
Formal (non free-zone) industry	Entered Ghanaian industrial scene in late 2009 Registered with the registrar general's department Mostly local/foreign partnership Operation based on market principles Export e-scrap openly and mostly added little value	Competition from informal sector operations State controls/ regulations governing scrap exports	Establish direct contact with waste pickers and refurbish shops Pre-finance informal collection Recruit reliable local agents Pay competitive prices with bonus

Informal local industry	- Not registered - Mostly family-based firms and small scale - Export e-scrap illegally in response to state regulation - Exports disguised as mixed scraps (ground) - Mainly local firms, but Nigerians also involved - Use rudimentary technology and old equipment - Low investment in buildings and equipment	- The challenge posed by their informal status and illegal operations - Criticisms from environmental NGOs - Lack state recognition - Stigmatisation - Harassment	- Use family labour and rudimentary tools - Target and negotiate with public institutions as well as importers of used electronics to buy bulk e-scraps - Recruit and entice (motivate) waste pickers - Pay competitive prices for products - Operate with low overhead cost - Adopt grinding to conceal product

Source: Fieldwork 2012

The free zone companies had to diversify into more capital-intensive ventures (i.e., production of lead ingots, iron rods), while some also widened their raw material supply base, moving beyond the boundaries of Ghana into Niger, Burkina Faso, Togo, etc. Others also resorted to pre-financing their agents, a practice not pursued prior to 2010. However, it remains to be seen whether knocking EPZ companies off their exclusive perch will bring about their demise, as local actors engage in fierce competition with them.

Parsing informal e-scrap collection

Though e-scrap collection takes place within different geographies and through different channels, "truck pushers" are the most consistently visible participants. They operate within neighbourhoods looking for recyclable and reusable materials and at times scavenge at the dump sites. Their operating in neighbourhoods comes at a cost, as the materials have to be paid for, but those who work at the dump sites collect these materials for free. With increasing competition, reliance on apprentices at mechanical and refurbishing workshops and house-helps in gated communities to separate and save recyclable materials for a negotiable fee has become common practice.

Though most truck pushers are aware of what materials recyclers prefer, at times collection is influenced by the availability of materials in their work zone. Some collect a diversity of materials based on their anticipated sale value. It is not unusual to encounter abandoned waste within the community; a practice frowned upon by scrap merchants and usually associated with occasional collectors with no community ties. Most survey respondents worked in the same zone every day, and indicated that the job is time intensive (on average, six days per week and nine hours per day). Most participants earn a daily average of 20 cedi per day.

A few primary collectors have the privilege of deciding what type of materials to collect based on fluctuations in market prices. These are experienced agents who operate from public institutions and offices and spend time preparing materials for resale by sorting or classifying them. In the face of increasing competition, they were, however, reluctant to reveal their "buyers" and "sellers." Generally, recyclers' selling behaviour is influenced by the distance between their operational turf and the Agbogbloshie (the e-scrap recycling hub in Accra), and the perceived economic returns.

Most truck pushers relied on one depot for selling their materials, but about 17 per cent indicated they sold at a location away from "home." About half the interviewees cited the prices offered, loyalty to clients who pre-finance their operations and the pay-as-you-go-method as the most important consideration in choosing a selling location, while about 20 per cent, especially those who work outside the Accra metropolis, are influenced by the location's convenience. Some recyclers accumulate in bulk before selling to scrap dealers, especially when they anticipate price hikes due to future shortages. Most recyclers, however, need cash immediately and cannot afford to speculate. In most cases, it is middlemen who sift through the scraps and separate them into different components – reusable copper, iron, etc. They have developed technical expertise in detecting the value of scrap based on its precious metal content, and the ability to gauge demand on the market.

Importantly, specialisation of activities in the e-scrap industry increases with each upward step on the recycling rung. According to Grant and Oteng-Ababio (2012), generalists operate at the base by sifting and hauling, while scrap dealers have varied and greater specialised knowledge about separation and market dynamics. This has two important implications: livelihoods can be precarious due to price fluctuations, and, more importantly, there can be large differences in earnings among those who collect the same materials in the same quantity, depending on when and where they sell. For example, those who sell to "convenience shops" in peri-urban areas earn less than those who sell to the scrap merchants of Agbogbloshie.

Closing the waste management loop

Recycling waste in confidence – local and free zone companies

The saying "one man's trash is another man's treasure" rings true of the world of solid waste management. One company exemplifying this reality is Special Steel Company Limited in Tema. Established in 2008, the company uses mixed scrap as feedstock for producing iron rods for local markets. The company, with a staff of about 150, procures on average about 45,000 tonnes of scrap per month locally at 500 cedi per tonne, and produces about 35,000 tonnes of iron rods monthly. The company has no raw material challenges, but the global economic downturn after 2008 negatively impacted the construction industry, leading invariably to decreased production levels. This situation was blamed on the "dollarisation" of the economy, with the fluctuations in the dollar rate creating unstable prices for the local economy. It also affected the export trade and the company's profit margins, and by extension the whole local e-scrap industry.

It was also noted that about 10 per cent of monthly e-scraps purchase contain high quantities of copper and are unsuitable for the company's purposes. Such "unwanted materials" are sold to other companies, which use copper as a major product component. Generally, recycling is a multi-step process, starting with collection and transportation of raw e-scrap, and progressing through pre-treatment, melting, refining, forming and finishing.

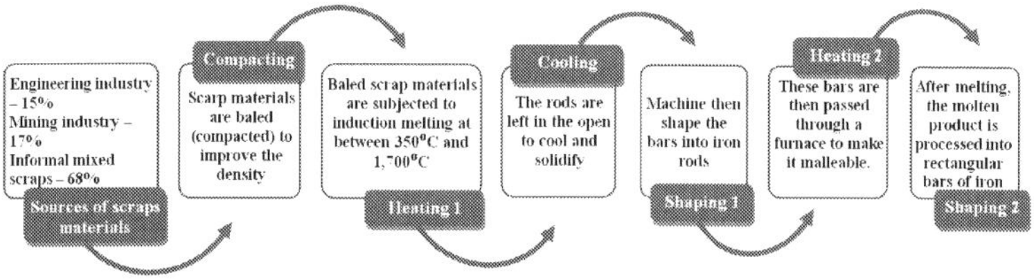

Figure 4.1:
Stages of turning e-scrap into iron rods
Source: Author's fieldwork (2012)

Figure 4.1 traces the stages in the production process for turning e-scrap into iron rods. Horizontally, the company operates with many accredited agents, who facilitate the company's linkages with the broader urban economy. These agents have the financial muscle to build bulk at the local level through their engagement with informal e-scrap collectors. To ensure client loyalty, the company at times pre-finances the collectors' activities through its trusted agents.

Linking the local and international economy

The general Ghanaian scrap economy is understudied, especially the export trade, which flies under the trade radar. This is largely because the scrap trade has been wrongly conceived as being at odds with Ghana's classic trade profile as an exporter of minerals, agricultural products and other natural resources. The scrap trade was conducted in secrecy and many players only operated on a need-to-know basis. Many people participated in just one part of the circuitry and within narrow ranges of business specialisation. Lack of knowledge and know-how, the inability of local entrepreneurs to develop business plans and failure to respond positively to the market mechanism accounted for e-scrap's going virtually unnoticed and being mis-documented until recently. Table 4.2 presents the exports of e-scrap from Ghana from 2004 to 2011.

Table 4.2: Key e-scrap exports from Ghana, 2004-11

Year	Copper	Mixed scrap	Lead	Totals
2004	1,866,548	13,163,567	1,806,519	16,848,638
2005	11,592,805	101,242,565	31,191,222	144,028,597
2006	59,866,623	60,694,752	58,913,797	179,477,178
2007	49,429,888	132,823,926	51,882,175	234,137,796
2008	155,516,588	109,175,523	202,472,016	467,166,135
2009	4,000,858	4,344,438	52,543,850	60,891,155
2010	NA	NA	109,559,883	109,561,893
2011	12,775,352	101,404	123,746,424	136,623,180
Total Metal	**295,048,662**	**422,555,975**	**632,115,886**	**1,348,734,572**

Source: Fieldwork (2012)

The national statistics for scrap exports show a rising export trend from 16,848,638 in 2004 to 136, 623, 180 in 2011, almost an eight-fold increase. Of particular significance is the mixed scrap category (powered scrap), a mis-description intended to outwit state officials tasked with enforcing the ban on ferrous metal exports. Mixed scrap exports rose in tandem with the ban, which remained in place from

2004 to 2008. As already indicated, the ban compelled local firms to adopt more diversified export strategies (misdescription, grinding) to remain competitive and active in the industry. Figure 4.2 provides an example of how a local company, operating through a parent company, Nord-Schrott Int, exports scraps from Ghana, with Germany and China as the main destinations.

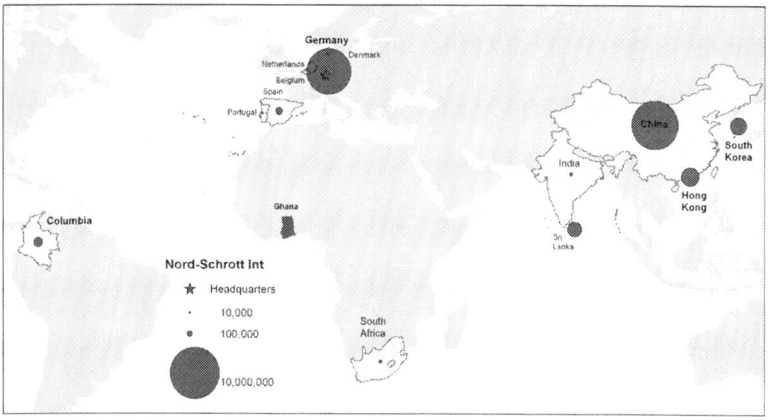

Figure 4.2:
Map showing Nord-Schrott Int. trading destinations from Ghana
Source: Author's fieldwork (2012)

The volume and weight of global e-scrap is expected to more than double in the next 15 years (Pike Research 2011). This will not be a simple North-South flow, as the available data clearly depict. Already, countries in the global South are producing one quarter of the world's e-scrap and they may overtake the richer countries as early as 2018 (*Economist* 2013). How Ghana positions itself in these global dynamics may impact not only the economy but also the quest for a liveable, zero waste city.

Cooperating for efficient environmental management

The way e-scrap is coordinated and governed has major implications for how formal and informal players access local and global markets and also for how they consolidate and perpetuate their operations within the local environment (Humphrey and Schmitz 2001). The formal sector complies with both internal and destination regulations and these are integrated (albeit weakly and inconsistently) into sectoral technical standards, as in the case of the free zone companies. The reality suggests that while the value chain for e-scrap may be buyer-driven, governance is exercised differently in different nodes along the chain.

This chapter shows that the success of the chain of activities has more to do with how the unrecognised and often harassed Kaya bola operate. Current relationships between formal companies and these primary collectors, often mediated by middlemen, are varied, due to their different capabilities. The alliance exhibits uncertainty in terms of consistency of supply, as the cost of switching between companies is reasonably low, making supply price-elastic and insecure. At the same time, all the partners have opportunities to ignore "contractual obligations" with impunity, since these are not tenable in any law court. Such a loosely regulated local alliance between formal and informal operators tends to perpetuate social inequality and increase environmental burdens.

The study results further suggest that attempts to distinguish between the formal and informal sectors within the scrap industry may be fruitless, and must be actively resisted, at least, at local authority level. The study points rather to a continuum between formality and informality (see Etzold et al. 2009), often masked by ignorance by downstream collectors and city authorities of the actual economic and environmental value of the "waste." Clearly, the strategy underlying Kaya bola operations appears to be gradualist and experimental, but the pace at which they link the formal and informal economy vertically and horizontally can be frenetic. For now, the Kaya bola remains key platform for institutional innovation, which authorities must encourage (and perhaps discard economically unsuccessful foreign-based models).

The challenge may be how to encourage public officials to buy into such a local model for pursuing environmental and social agendas, with the added advantage of reducing poverty. However, a local alliance of bureaucrats and "informalists" is unlikely to materialise, not least, perhaps, because the greater share of environmental burdens and social inequalities are borne by more distant populations or disadvantaged groups. Clearly, reforming urban governance while maintaining healthy urban economies will require new forms of experimentation. From that perspective, the locally driven Kaya bola model that has been successful in mitigating inequality and growth-related environmental burdens cannot be simply ignored. Reforming urban governance is likely to be crucial.

Conclusion

This chapter explores the impact of the breakthrough zero waste city concept in revolutionising municipal solid waste management in a developing country and identifies the Kaya bola as a possible conduit for addressing urban inequality and growth-related environmental impacts. The zero waste philosophy aims to minimise waste and resource consumption to conserve energy, mitigate climate change, reduce water usage, prevent the creation of toxins and minimise ecosystem destruction. This is against the backdrop that most cities in the sub-region, including Accra, face enduring solid waste management challenges, with no regulatory frameworks to push sustainable policies, limited capacity and the weakest drivers of environmental sustainability, all compounded by institutional poverty (UN Habitat 2012; Okot-Okumu and Nyenje 2011).

As city authorities and businesses commit to proper environmental management, they tend to adopt "game-changing" innovative techniques (mostly foreign-based), which in most cases are pregnant with the risk of failure (Oteng-Ababio et al. 2013). This study amply demonstrates that from source reduction to eliminating incentives for raw material extraction, the path to zero waste requires moving beyond the often theoretical rhetoric of recycling to a more practically based integrated solid waste management system (Scheinberg et al. 2011). Crucially, such a holistic system must guard against the continued social and economic marginalisation of sections of the population, such as the informal recyclers, and encourage waste separation that provides cleaner feedstock to support the vibrant e-scrap industry (RICS 2011).

Accra's journey towards becoming a zero-waste city may be long and arduous, complex and expensive, but must be completed if the city is to live up to its millennium city status. Clearly, traditional, deep-seated urban frameworks for understanding the city and the informal sector have no way of comprehending and appreciating the role of the Kaya bola in this drive. Currently, they battle doggedly to keep livelihoods alive just as the government battles determinedly to throw them out of business. Yet, the Kaya bola sector has amply demonstrated it can combine alternative modes of survival, work and collective action that often render entrepreneurship a form of business activism (Medina 2005; Obeng-Odoom 2011). That notwithstanding, Kaya bolas continue to operate in hostile social environments and are often treated as a nuisance, embarrassment or as criminal. Thus, they largely remain unorganised, unrepresented and unprotected.

We need to move away from seeing waste collection by the informal sector in terms of survival livelihoods to conceptualising this process within a broader economic sustainability context. This is an environment-livelihood nexus that offers considerable possibilities for creating green jobs (Grant and Oteng-Ababio 2012) and also for creating a platform in which the "global" and the "local" interact vigorously. The informal e-scrap economy provides a glimmer of hope and a creative pathway for what could be achieved globally, the zero waste city. There is no clear path, but a promising way forward would be to encourage all the stakeholders to reconfigure the formal-informal interface. Admittedly, such a venture is likely to be more expensive for the municipality and so is unlikely, yet literature is replete with "abundant grace" for all cities in developing countries that have successfully walked this path (Medina 2005; Sternberg 2013).

References

Adama, O. (2012). Urban governance and spatial inequality in service delivery: A case study of solid waste management in Abuja, Nigeria. *Waste Management Research* **30**(9): 991 – 8.

Amankwaa E.F. (2013). Livelihoods in risk: exploring health and environmental implications of e-waste recycling as a livelihood strategy in Ghana. *The Journal of Modern African Studies* **51**(4), 551 – 75.

Aryeetey E. and F. Tarp (2000). Structural Adjustment and After: Which Way Forward? In *Economic Reforms in Ghana, The Miracle and The Mirage*, E. Aryeetey, J. Harrigan and M. Nissanke (eds). Oxford: James Currey and Woeli Publishers.

Coe, N. M., P. Dicken, and M. Hess., (2008) Global production networks: Realizing the Potential, *Journal of Economic Geography,* **8:** 271 – 95.

Etzold, B., M. Keck, H. G., Bohle, and W. P. Zingel (2009).Informality as agency – negotiating food security in Dhaka. *Die Erde*, **140**(1): 3-24.

Fahmi, W.S. (2005). The impact of privatization of solid waste management on the Zabaleen garbage collection of Cairo. *Environment and Urbanisation* **17**(2): 155 – 70.

Grant, R. (2009). *Globalizing City: The urban and economic transformation of Accra*. Ghana/New York: Syracuse University Press.

Grant, R. and M. Oteng-Ababio (2012).Mapping the invisible and real African economy: Urban e-waste circuitry. *Urban Geography* **33**(1): 1 – 21.

Humphrey, J. and H. Schmitz.(2001) Governance of global value chains.IDS Bulletin, 32, 3. Institute of Development Studies

McGranahan, G. and G. Martine (2012). Urbanisation and development: Policy lessons from the BRICS experience. Human settlements discussion paper. London: IIED.

Medina, M. (2005).Serving the unserved: Informal refuse collection in Mexican cities. *Waste Management and Research* **23**(5): 390 – 7.

Wes Muir, M. (2009) Footprints in waste management: Taking steps toward zero waste, *Tree Hugger,* http://www.treehugger.com/corporate-responsibility/footprints-in-waste-management-taking-steps-toward-zero-waste.html Accessed on 13 June, 2013

Nzeadibe, T.C. (2009). Solid waste reforms and informal recycling in Enugu urban area, Nigeria. *Habitat International* **33**: 93 – 9.

Obirih-Opareh, N. and J. Post (2002). Quality assessment of public and private modes of solid waste collection in Accra. *Habitat International***28**(1): pp. 95–112.

Obeng-Odoom, F. (2011).The informal sector in Ghana under siege. *Journal of Development Society* **27**(3 and 4): 355 – 92.

Okot-Okumu, J. and R. Nyenje (2011). Municipal solid waste management under decentralisation in Uganda. *Habitat International* **35**: 537 – 43.

Oteng-Ababio, M. (2010). E-waste: An emerging challenge for solid waste management in Ghana. *International Development Planning Review* **32**(2): 191 – 206.

Oteng-Ababio, M. (2011a).The role of the informal sector in solid waste management in the GAMA, Ghana.*Tijdschrift voor Economische en Sociale Geografie* **103**(4): 412 – 25.

Oteng-Ababio, M. (2011b). Missing links in solid waste management in the Greater Accra Metropolitan Area in Ghana. *GeoJournal* **76**: 551 – 60.

Oteng-Ababio, M. (2012).When necessity begets ingenuity: Scavenging for survival in a globalizing city. *African Studies Quarterly* **13**(1 and2): 1 – 21. Spring. http://www.africa.ufl.edu/asq/v13/v13i1–2a1.pdf

Oteng-Ababio, M., E.J.A. Melara and G. Offira (2013). Solid waste management in African cities: Sorting the facts from the fads in Accra, Ghana. *Habitat International* **39**: 96 – 104

Parizeau, K. (2013). Formalization beckons: A baseline of informal recycling work in Buenos Aires, 2007 – 2011. *Environment and Urbanisation* **25**: 501 – 21.

Pike Research (2011). *Electronics recycling and e-waste issues: Reuse, recycling, storage, and waste management for consumer electronics, computer equipment and mobile phones: market analysis and forecasts.* Boulder: Pike Research.

Royal Institution of Chartered Surveyors (RICS) (2011).*Waste and the Built environment: The UK case study.* London: RICS.

Scheinberg, A., S.S. Simpson, H. Michael and A.P.J. Mol (2011). Assessing urban recycling in low and middle income countries: Building on modernised mixtures. *Habitat International* **35**: 188 – 98.

Sternberg, C.A. (2013). From "cartoneros" to "recolectores urbanos": the changing rhetoric and urban waste management policies in neoliberal Buenos Aires. *Geoforum* **48:** 187 – 95.

Songsore, J. (2010). The Urban Transition in Ghana: Urbanisation, National Development and Poverty Reduction. http://pubs.iied.org/pdfs/G02540.pdf. Accessed on 4 June 2012.

UN Habitat (2010). *Solid waste management in the world's cities: Water and sanitation in the world's cities.* London and Washington DC: United Nations Human Settlements Programme/ Earthscan Publications.

Van Hear, N. (1998). *New diasporas: The mass exodus, dispersal and regrouping of migrant communities.* London and Seattle: University College London Press and University of Washington Press.

Zaman, A. and S. Lehman (2011). What is the "zero waste city" concept?

http://w3. unisa. eduAu/artarchitecturedesign/ZeroWasteSAResearchCentre/docs/ZWC% 20Concept. Pdf. Accessed on Assessed on 13 June 2013

Zaman, A.U. and S. Lehmann (2013). The zero waste index: A performance measurement tool for waste management systems in a "zero waste city". *Journal of Cleaner Production* **50**: 123 – 32.

5

INFORMAL RECYCLING MARKET LINKAGES AND URBAN GOVERNANCE IN NIGERIA

Thaddeus Chidi Nzeadibe, Raymond N.C. Anyadike and Onyanta Adama

Introduction

Solid waste management in Nigerian cities is characterised by the lack of a formal resource recovery programme. As a result, the only form of public recycling is that undertaken by the informal sector, comprising waste pickers, scrap dealers, itinerant waste buyers, cart pushers, middlemen and micro and small enterprises. Waste pickers and scrap dealers are the main drivers of informal recycling in Nigeria (Nzeadibe and Ajaero 2011). The working definition of "recycling" adopted in this chapter is the recovery of materials by informal sector groups, and their valorisation, sale or use (Nzeadibe and Ajaero 2010: 245). Valorisation refers to "the entire process of extracting, storing, collecting, or processing materials from the waste stream in order to extract and divert value and direct the material to a value-added stream" (Scheinberg et al. 2010: 216).

Waste picking is the manual sorting and picking of recyclable/reusable materials from mixed wastes at or in landfills, dump sites, street bins and piles, transfer points as well as waste collection trucks (Anschütz et al.2004: 3).Waste picking is known to be critical in the recycling systems of developing country cities (Wilson et al. 2006; Medina 2007; Scheinberg et al. 2010) with over 15 million people or 1 per cent of the developing world's urban population estimated to be involved in it (Medina 2007: vii).

The political economy of waste recycling by the informal sector in Nigeria deserves mention here. In this context, Nzeadibe and Anyadike (2010: 1284) observe that the less than stellar performance of the solid waste collection, recycling and disposal system in Nigerian cities is attributable to governance problems and also a tangible shortfall of financial resources for local government. As a result, municipal governments are unable to provide these urban environmental services on a sustainable basis (Egbu et al. 2012), thereby contributing to the service deficits in the waste sector. Interestingly, the existence of deficits in solid waste management services has often created opportunities for informal actors in urban Nigeria. Hence, the most visible and prevalent form of public recycling is that undertaken by the informal sector.

On the other hand, one of the understudied geographies of the world economy is the pattern and material linkages in the informal waste economy of the global South (Brooks 2012, 2013). In Africa, informal transborder trade in recyclable materials has received little or no attention. Althougha range of actors is known to operate in the informal waste value chain,so much remains unknown about the actors in this sector, their activities and their lives (Nzeadibe et al. 2012).

In addition, research into the spatio-organisational patterns of informal recycling in urban Nigeria has been limited (Nzeadibe and Eziuzor 2006; Nzeadibe and Iwuoha 2008), while regional linkages in the markets for recyclables and the implications of observed patterns for urban development appear not to have been considered. The local and sub-regional material supply chains and linkages in the recycling economy have also received little attention from researchers, leaving us with incomplete knowledge of this dynamic system (Medina 2011). The present chapter focuses on waste pickers and scrap dealers, because they are arguably the main drivers of informal recycling in Nigerian cities.

Justification for the study

To better comprehend the nature of dynamic informal economic activity and to contextualise the subject in a spatial perspective, this chapter focuses on the city of Aba, in southeastern Nigeria known for its burgeoning informal economy(Meagher 2009, 2010 and 2011). Recent estimates of other segments of the informal urban economy in Aba, namely the shoe- and garment-making sectors, indicate that together they employed over 58,000 people and had a combined annual turnover of over US$200 million (Meagher 2009,2010). Unfortunately, comparative data on the recycling sector of the waste economy is not available. As a result, the present chapter relies on estimates and observations made during our field survey.

This chapter seeks to situate informal recycling activities in an urban and regional context in Nigeria with a view to highlighting their potential contribution to the urban economy and their implications for urban development policy in Nigeria.

Study area

Aba is the commercial and industrial nerve centre of Abia State, southeastern Nigeria. This city provides goods and services to adjacent urban areas and international markets (Nwafor 2002). It is located between latitudes 5°2'N and 5°11'N, and longitudes 7°18'E and 7°28'E (Figure 1). The study area covers an estimated 95 km^2 (Uchegbu 2009). It includes the entire Aba North and Aba South Local Government Areas (LGAs) and parts of Osisioma and Ugwunagbo LGAs (Figure 5.1).

The population of Aba has been growing steadily since the first census of the study area in 1931. Between the 1931 and 1952/53 censuses, the population of Aba increased fourfold from 13,000 to 58,000 (Okoye 1975, 2002). By 1953, Aba was classified among the urban centres in Eastern Nigeria with 50,000 or more inhabitants (East Central State of Nigeria 1972; Okoye 1975).In 1963, the population was 131,003 and ten years later180,558 (Okorafor 1975). By 1991, the population of Aba had risen to 494,152 (National Population Commission 1991). As of 2006, the estimated population stood at 1,022,138 (Federal Republic of Nigeria 2009).

Aba is also a rapidly growing city in terms of concentration of socioeconomic activities and general urban expansion. Its urban growth and development has been largely unplanned, resulting in various incompatible land uses in close proximity. With residential, industrial and commercial land juxtaposed, urban environmental problems such as noise, traffic congestion, air pollution, street hawking and solid waste are rife.

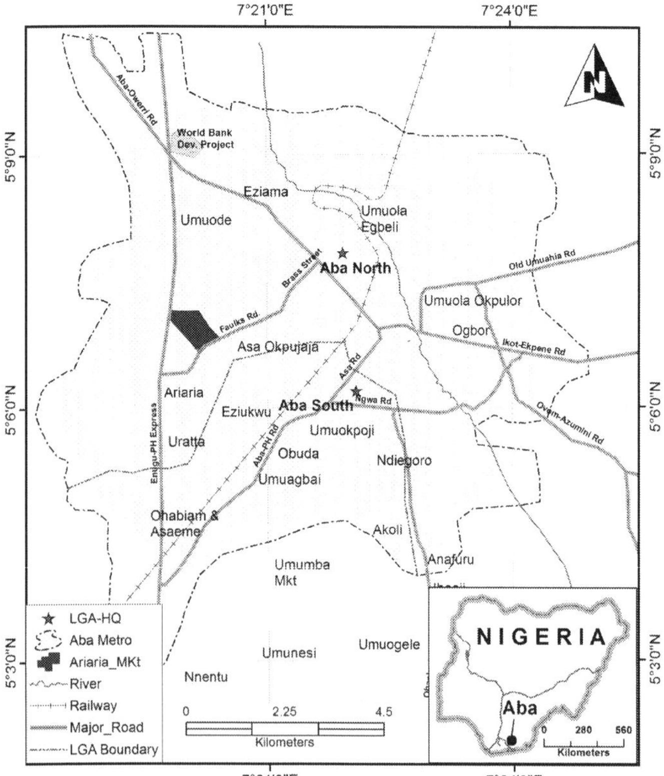

Figure 5.1:
Map of Aba urban area, Nigeria

Research Methodology

A mixed methods approach was adopted for this research. It comprised primary data collection using Key-Informant Interviews (KIIs), field observation and a survey of 401 waste pickers and 195 waste dealers. Local and international literature on waste recycling by the informal sector constituted the secondary data sources.

In order to investigate the linkages in the Aba recycling economy, KIIs were conducted with four major waste dealers and the zonal leader,Iron and Bottle Dealers' Association (a sub-set of the Old Court Main Market, Aba) and the president general of the Old Court Main Market, Aba. KIIs by telephone and email were also conducted with two businessmen involved in the sale and transport of scrap materials to Aba from Kumba,Southwest Province,Republic of Cameroon. Kumba, with a population of about 120,000, is one of the most economically vibrant towns in southwest Cameroon. It is a meeting point of traders from Nigeria, Gabon and Equatorial Guinea, who do businesses in the southwest region (Ngoasong 2007; Asong 2010). It is the largest town in the region and is characterised by rapid economic growth, especially among small and medium enterprises, an possesses a dynamic informal economy (Acho-Chi 2002; Ngoasong 2007).

Additionally, KIIs were held with a professor with extensive research experience of and publications on solid waste management, and four researchers in the University of Nigeria's faculties of physical science, agriculture and social sciences who were born and raised in Cameroon and are familiar with informal recycling activities. These interviews yield further insights into the issues in scrap recycling raised by the two businessmen.They also enabled us to determine the importance of the informal recycling sector in terms of its linkages tothe international recycling market and the formal sector.

Results and Discussion

Waste picking in Aba is carried out at urban refuse dumps and along the streets. Municipal authorities collect household refuse for disposal at designated dumps. Residents and businesses also dispose of their waste at government-approved dump sites. Beyond collection and dumping, however, little effort is made by local government refuse crews to separate or recover recyclable materials. It is therefore through the efforts of the waste pickers that material recovery and separation of recyclables is achieved. It was also observed in the study area that, just as in other Nigerian cities (Agunwamba 2003), no formal resource recovery and recycling programmes exist. One can safely assert that the recycling of wastes in Aba depends on the efforts of waste pickers.

Outlets for Recyclables and Linkages with Local Formal Economy

Major outlets for the sale of recyclable materials in Aba are shown in Table 5.1. This table indicates that more than half the recyclables recovered and traded by the informal sector in Aba are consumed by formal sector companies, while more than a quarter are used by small- and medium-scale enterprises.

Table 5.1:
Percentage distribution by major outlet of sales of recyclable materials

Outlets	Per cent
Companies	55.4
Small and medium enterprises	27.2
Cooperatives	17.4

Source: compiled by the authors

It was found that certain local industries in Aba rely on supplies of secondary raw materials from waste pickers and waste dealers. For example, KIIs with some waste dealers revealed that they provided the beverage industry (e.g., Nigerian Breweries and Nigerian Bottling Company), paper industry (e.g., Star Paper Mill), glass industry (e.g., International Glass Industry) and pharmaceutical industry with reusable/recyclable materials such as glass bottles and waste paper.

One scrap dealer remarked: "Obasanjo's government policy had major impact on scrap business. This led Aba recyclers to export processed scrap metals to Lagos and Ibadan for further processing in companies [formal]. Some Indians and White men come to Aba to buy scraps" (Interview with key informant on 10thNovember 2010). This clearly shows that the informal recycling sector in Aba is well articulated with the formal economy in terms of materials supplied and used. The sector acts as a supply chain for secondary raw materials to industry. This finding accords with those of Medina (2000, 2007) concerning the linkages between the informal and formal sectors in Asia and Latin America.

Socio-Spatial Organisation of Aba's Recyclable Markets

Meagher (2010) distinguishes between informal governance processes, which relate to the regulatory role of social networks, informal institutions and popular associations in urban areas, and formal governance processes, which involve formal political, economic and social institutions. In this context, the organisation of the powerful, tiered waste dealers' association, comprising individuals involved in the waste trade in Aba, may be regarded as informal urban governance. Whatever materials the waste pickers salvage must be sold for either direct reuse or recycling in order to complete the recycling loop. The major market clusters for recyclable materials identified during the study are Old Court Main Market, School Road (the largest); Cemetery/Eziukwu Market; Over-rail Market; Uratta Market; Ariaria Main Market; and Ekeoha Market.

Although informal recycling is popularly perceived as "an unorganised activity," and remarkably little is known about how people in informal recycling organise themselves at work in Nigeria (Nzeadibe and Iwuoha 2008; Adama 2012), there seems to be some order or, more appropriately, socio-political organisation among the waste dealers in Aba. However, most of the individual waste pickers operate solo, do not have viable cooperatives and have limited negotiating power in their transactions with middlemen. Thus, waste pickers often find themselves in exploitative relationships with scrap dealers.

By contrast, the other end of recycling, namely the buying and selling of recyclable materials and the market for them, is highly organised. Market unions for the various categories of material exist and they have a very influential association with a powerful chairman and executive committee. The market union exercises considerable sway over the prices paid for materials from waste pickers or sold to end users or industry.

The markets for recyclables in Aba are also well organised. For example, Old Court Main Market comprises five zonal unions on the basis of the wares sold, the most important being glass bottles, plastics and metal scraps (Interview, zonal leader of Iron and Bottles Dealers, Old Court Main Market, Aba, 5 March 2006).Each zone has a leader, while a president general of the entire market union oversees the activities of all the zonal unions. The president general, in consultation with his executive committee, is the only person who can authorise interviews or research of any type in the market. The unions also seek to influence urban policy and law enforcement agents such as the police, preventing them from unduly harassing their members at the markets. This organisation of the recyclables market in Aba suggests that certain of its sub-sectors have achieved significant development and stability.

Processing and sale of materials

Recovered materials are aggregated and weighed during processing for sale to middlemen. To recover metal embedded in plastic materials such as battery casings and tyres, the materials are burned in order to expose the metals. Processing of recovered recyclables also includes size reduction, washing and cleaning, packaging, etc. These operations are "value-addition measures" intended to increase the market value of the wastes or to facilitate their handling and transportation.

Some waste pickers are also entrepreneurs, buying scrap materials from colleagues for sale to higher-tier middlemen. Middlemen act as a link between the waste pickers and industry. The latter requires a minimum quantity and quality of materials, and only the middlemen can guarantee both. No industry will buy secondary raw materials directly from individual pickers, because it would not be economic. Consequently waste pickers are vulnerable to exploitation by middlemen.

Transportation of Recovered Materials

Most of the waste pickers (80 per cent) indicated that the dominant form of transport for recovered materials is trekking on foot and human porterage. A few waste pickers, however, use push carts to transport their wares to market or to the waste dealers who patronise them. This mode of transport eats into valuable picking time (Agunwamba 2003).Consequently, Agunwamba (2011) has argued for development intervention in the form of tricycles to waste pickers to improve their efficiency at work. On the other hand, waste dealers employ various means to transport their wares. For example, small waste dealers are known to use push carts to freight their goods locally. They often use pickup trucks and vans to convey recyclables from the nearby cities of Port Harcourt, Calabar and Uyo. Major dealers employ heavy trucks (popularly called 911 trucks) for goods coming from outside Aba or going to the cities of Onitsha, Warri and Lagos. Push carts and pickup vans may be used for the local transport of wares by small dealers.

Regional Material Linkages in the Aba Recycling Economy

Growing awareness of the size and organisational potential of informal economies in an era of globalisation has triggered a profusion of research into the linkages between the formal and informal economies (Meagher 2013). An attempt was made for the purposes of this chapter to determine the origin of the buyers of recyclable materials in Aba. Figure 2 presents the states of origin of such buyers in percentages. It reveals that most of the buyers (45.5 per cent) are of local origin, operating within Aba as artisans, small and medium enterprises, cooperatives or companies. The items bought are used for a variety of purposes, such as building, car repairs and household use. Other major places of buyer origin are Onitsha in Anambra State (17 per cent) and Lagos (16.8 per cent).

It should also be noted that due to proximity, the industrial economy of Port Harcourt, Rivers State, acquires 5.8 per cent of recyclables from Aba. Regionally, buyers originate from three Southern Nigerian geopolitical regions, namely southeast, southwest and south-south. Figure 5.2 also shows the relative importance of regional material linkages within the Aba recycling economy.

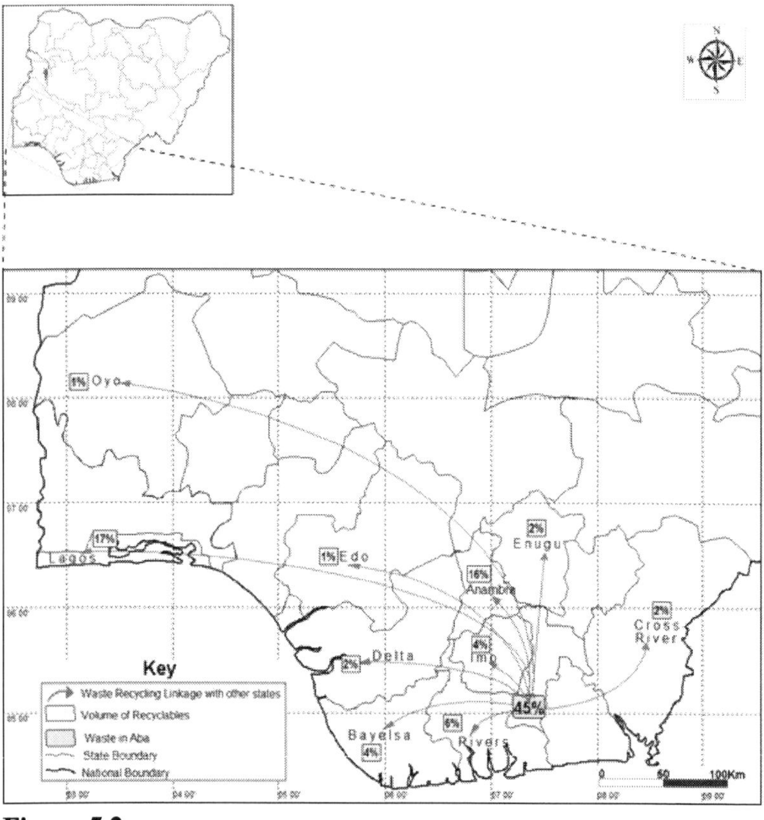

Figure 5.2:
Regional material linkages in the Aba recycling economy
Source: Author's fieldwork 2010

Trans-border Trade in Scrap Materials

Global structures and processes of change are having significant impacts on the recycling economy in many countries of the world

(Medina 2007, 2011). In response to growing global demand for recyclables and the resultant increase in transborder trade, the internationalisation of recycling networks between the commercial city of Aba and neighbouring countries was observed.

A dynamic recycling economy involving scrap dealers links the cities of Malabo in Equitorial Guinea, and Kumba and Douala in Cameroon to the city of Aba in southeastern Nigeria. The waste dealers freight in materials from Malabo and Cameroon by barge, boat and road. Upon arrival at the ports of Calabar and Oron, the materials are hauled over land to Aba using trailers and heavy trucks. The following information is derived mainly from our interviews with Cameroon-based scrap dealers.

> Scrap materials traded between Cameroon and Nigeria includes metals like copper, aluminium and iron as well as plastics and glass materials. Metals are melted for other uses like aluminium for kitchen utensils. One kg sells for 1500 CFA 0r ₦506[based on an exchange rate of ₦ 1 to CFA 2.94 obtained on 5 April 2011]. One kg of iron sells for 100 CFA [₦ 34].Copper sells for 2,800 CFA [₦ 945]. When these are recycled, they are used to produce building rods and flat sheet for the repair of vehicles (Email Interview with Collins Akukwe 4th April 2011).

The KIIs also reveal that:

> Some bottles like that of glycerine are bought, rinsed and reused in the refilling of the product. Its cost ranges from 50 CFA [₦ 17] to 100 CFA [₦ 34]. Also, there are the car batteries that are sold as scrap and the [price] range is about 170,000 CFA [₦57,352] per ton. The amount of metals sold in a month varies (Email Interview with Collins Akukwe 4th April 2011).

> Sometimes about 1,500 tons of metals are sold per month depending on the number or frequency of people who come to buy it. Metal business is lucrative in Kumba and Southwest Province. As such, the government of Cameroon benefits from it through the collection of taxes and also its provision of jobs to the economy. For the factors encouraging the trade, means of transport is positive from where the materials are found to the market. The major means of transportation is by road (Email Interview with Collins Akukwe 4th April 2011).

Corruption is inevitable in all societies. Police officers sometimes come up with charges like disrespect for traffic rules and allegations of overloading the vehicles. By so doing, they extort money from the scrap dealers coming from Nigeria. The business climate between Cameroon and Nigeria is generally friendly (Email Interview with Collins Akukwe 4th April 2011).

This last fact appears to be confirmed by recent research into the Cameroonian waste recycling system, which asserts that Nigerian businessmen involved in recycling had active links with intermediary agents in Cameroon (Asong 2010). It would also seem that this international linkage has existed for over ten years. This view is supported by Asong (2010) who asserts that "after 2000, there has been a growing market for aluminium and other metal wastes, some of which are transported abroad to countries like China, Ghana and Nigeria" (pp.95 – 6). The scraps are sold either to local private recycling services or to neighbouring Nigeria (p.102).

This view was recently echoed by Mbeng (2013: 478), who noted in a study of informal waste recovery in Douala, Cameroon that "from interviews, only a small proportion of plastics and scrap metals are recycled in Douala, with a greater proportion being exported to Nigeria, endowed with a bigger market."

As one waste dealer in Aba put it during his interview:"We travel to Cameroon and Malabo to buy scrap metals such as aluminium, copper and brass and ship them to Nigeria through Calabar and Oron ports. The people of Malabo don't know about scraps" (Interview, zonal leader of Iron and Bottles Dealers, Old Court Main Market, Aba, 5 March 2006).

Conclusion and Policy Implications

This chapter has shown that informal waste recycling in Aba very largely depends on the efforts of waste pickers who daily comb the urban dumps in search of reusable and recyclable materials, and waste dealers who trade in recovered materials. Waste picking in Aba attempts to fill the gap left by the lack of formal recycling programme. Informal recycling is therefore an important but neglected aspect of waste and resource management in Nigeria. The chapter draws attention to aspects of informal recycling that may be useful in the formulation of public policy on solid waste management, for example, in the planning and routing of waste collection and recycling. Where such policy recognises waste pickers as stakeholders in the solid waste management process and as important actors in urban economic space, then a more sustainable and inclusive management process in Aba is possible. In addition, this chapter indicates the flow pattern of recyclable materials in Aba and beyond. The recent completion of the Nigeria-Cameroon highway and the associated transport facilitation programme can play a key role in the movement of recyclables between the two countries, thereby boosting regional trade and integration.

Although our research did not aim to quantify the material flows within the system, the evidence suggests the dynamic informal waste recycling economy in Aba is closely integrated into both the formal economy of Aba and Nigeria and the international circuitry of the waste economy. Thus, informal recycling in Aba has strong forward and backward linkages with the local formal and the international economy, and the waste sector ought to be recognised as making a substantial contribution to national income.

This chapter echoes earlier findings elsewhere in Nigeria that the informal waste sector possesses enormous development potential if properly recognised and supported (Nzeadibe and Ajaero 2011).Unfortunately, this development contribution is not well known to governments and citizens in Nigeria. As a result, informal waste operators have been the victims of hostile urban policies and social opprobrium and informal waste management has gone unrecognised in urban and environmental planning (Nzeadibe et al. 2012).

A necessary first step in reforming the solid waste management system is official recognition of the informal sector by those in authority and acknowledgement of the benefits of informal recycling. It must be emphasised that recycling should not be left in the hands of waste pickers and scrap dealers alone. If they are serious about reducing poverty in Nigeria, governments should get involved by supporting informal sector groups and cooperatives through appropriate economic policies and by implementing recycling programmes.

References

Adama, O. (2012). Urban livelihoods and social networks: Emerging relations in informal recycling in Kaduna, Nigeria. *Urban Forum* **23**(4): 449 – 66. DOI:10.1007/s12132-012-9159-8

Adeyemi, A.S., J.F. Olorunfemi and T.O. Adewoye (2001).Waste scavenging in Third World cities: A case study in Ilorin, Nigeria. *Environmentalist* **21**(2): 93 – 6.

Acho-Chi, C.(2002).The mobile street food service practice in the urban economy of Kumba, Cameroon. *Singapore Journal of Tropical Geography* **23**(2): 131–48.

Asong, F.Z.(2010).Recycling and material recovery in Cameroon: Implications for poverty alleviation and ecological sustainability. Unpublished PhD dissertation, Brandenburg University of Technology, Cottbus, Germany.

Agunwamba, J.C. (2003).Analysis of Scavengers' Activities and Recycling in some cities of Nigeria. *Environmental Management* **32**(1): 116 – 27.

Agunwamba,J.C.(2011). Key-informant interview with renowned researcher on solid waste management in Nigeria and professor of engineering, University of Nigeria, 17 January.

Anschütz, J.,A. Scheinberg and A. van de Klundert (2004). *Addressing the Exploitation of Children in Scavenging (Waste Picking): A Thematic Evaluation on Action on Child Labour*. Geneva: International Labour Organisation.

Brook, A. (2012). Riches from rags or persistent poverty? The working lives of second-hand Clothing vendors in Maputo, Mozambique. *Textile: The Journal of Cloth and Culture* **10**(2): 222 – 37.**DOI:** http://dx.doi.org/10.2752/175183512X13315695424239

Brooks, A. (2013).Stretching global production networks: The international second-hand clothing trade. *Geoforum* **44**: 10 – 22. DOI http://dx.doi.org/10.1016/j.geoforum.2012.06.004

East-Central State of Nigeria (1972).Statistical Digest 1970.Official Document No. 8.Enugu: ECS.

Egbu A.U., E.C. Mbano-Pat and K.A. Obialo (2012).Public goods provision: Informal response to government failure in the cities of Nigeria. *Theoretical and Empirical Researches in Urban Management* **7**(2): 67 – 78.

Federal Republic of Nigeria (2009).Official Gazette No. 2, vol. 96. Abuja: Federal Government Printer.

Gunsilius, E. and A. Whiteman (2010). Waste management and governance: Collaborative approaches for public service delivery. In *Solid waste management in the world's cities*, A. Scheinberg, D.C. Wilson and L. Rodic (eds).London: Earthscan.

Mbeng, L.O. (2013). Informal Waste Recovery and Recycling: Alleviating Poverty, Environmental Pollution and Unemployment in Douala, Cameroon.*Journal of Scientific Research and Reports* **2**(1): 474–90.

Meagher, K. (2009): Trading on faith: Religious movements and informal economic governance in Nigeria. *Journal of Modern African Studies* **47**(3): 397–423.

Meagher, K.(2010). *Identity economics: Social networks and the informal economy in Nigeria.* New York: James Currey.

Meagher, K. (2011).Informal Economies and Urban Governance in Nigeria: Popular Empowerment or Political Exclusion? *African Studies Review* **54**(2): 47–72.

Meagher, K.(2013). Unlocking the Informal Economy: A Literature Review on Linkages Between Formal and Informal Economies in Developing Countries. WIEGO Working Paper No. 27. Women in Informal Employment: Globalizing and Organizing. Cambridge MA: WIEGO.

Medina, M. (2000). Scavenger cooperatives in Asia and Latin America. *Resources, Conservation and Recycling* **31**(1): 51–69.

Medina,M. (2007). *The world's scavengers: Salvaging for sustainable consumption and production*. Lanham MD: AltaMira Press.

Medina, M. (2011). Global Supply Chains in Chinese Industrialisation: Impact on Waste Scavenging in Developing Countries. Working Paper No. 2011/78.Helsinki: UNU-WIDER.

National Population Commission (1991). *Nigeria Population Census 1991*. Abuja: NPC.

Ngoasong, M.Z. (2007). Does Innovation Matter for Economic Development? An empirical study of small and medium-sized enterprises in the city of Kumba, Cameroon. *Negotia* **52**(1): 91–110. Retrieved from http://www.studia.ubbcluj.ro/download/pdf/300.pdf#page=92 on 13 April 2011.

Nwafor, J.C. (2002). Manufacturing industries. In *A survey of the Igbo Nation,* G.E.K. Ofomata(ed.).Onitsha: Africana First.

Nzeadibe, T.C. (2009). Solid waste reforms and informal recycling in Enugu urban area, Nigeria, *Habitat International* **33**(1): 93 – 9.

Nzeadibe, T.C. and O.J. Eziuzor(2006). Waste Scavenging and Recycling in Onitsha Urban Area, Nigeria. *CIWM Scientific and Technical Review* **7**(1): 26–31.

Nzeadibe T.C.and H.C.Iwuoha (2008). Informal waste recycling in Lagos, Nigeria. *Communications in Waste and Resource Management* **9**(1):24–30.

Nzeadibe T.C. and C.K. Ajaero (2010). Informal waste recycling and urban governance in Nigeria: Some experiences and policy implications. In *Handbook of Environmental Policy,* J. Meijer and A. der Berg (eds). New York: Nova.

Nzeadibe T.C. and R.N.C. Anyadike (2010). Solid waste governance innovations: An appraisal of recent developments in the informal sector niche in urban Nigeria. *Geography Compass* **4**(9): 1284–96.

Nzeadibe T.C. and C.K. Ajaero (2011).Development impact of advocacy initiatives in solid waste management in Nigeria. *Environment, Development and Sustainability* **13**(1): 163–77.

Nzeadibe, T.C., R.N.C. Anyadike and R.F. Njoku-Tony (2012). A mixed methods approach to vulnerability and quality of life assessment of waste picking in urban Nigeria. *Applied Research in Quality of Life* 7(4), 351-370.DOI: 10.1007/s11482-012-9171-0.

Okorafor, A.E. (1975).Population distribution (1963).In *Nigeria in-maps: Eastern states*, G.E.K. Ofomata (ed.). Benin City: Ethiope Publishers.

Okoye, T.O. (1975).The modern phase of urbanisation.In *Nigeria in maps: Eastern states*, G.E.K. Ofomata (ed.). Benin City: Ethiope Publishers.

Okoye, T.O. (2002).Urban life and urban development in Igboland. In *A survey of the Igbo Nation,* G.E.K.Ofomata (ed.). Onitsha: Africana First.

Scheinberg, A., D.C.Wilson and L. Rodic (2010).*Solid waste management in the world's cities*. London: Earthscan.

Uchegbu, S.N. (2009). Effective planning and management as critical factors in urban water supply and management in Umuahia and Aba, Abia State, Nigeria. *Physics and Chemistry of the Earth* **34**: 23–7.

Wilson, D.C., C. Velis and C. Cheeseman (2006).Role of informal sector recycling in waste management in developing countries. *Habitat International* **30**(4): 797–808.

PART THREE
SOCIAL DIMENSION

6

DYNAMICS OF INFORMAL SECTOR SOLID WASTE RECYCLING IN ZARIA, NIGERIA

Joy Ede Ukoje

Introduction

The informal sector has long been recognised for its potential for employment generation and growth (De Soto 1989; Sethuraman 1981). In Africa, the economic downturn that followed the introduction of the Structural Adjustment Programme is responsible for the changed conditions under which urban dwellers make a living (Meagher and Yunusa 1996; Ahonsi 2000). In the face of rapid urban growth, a generally low-skilled labour force and poor availability of wage jobs, urbanites now increasingly rely on informal sources of income. Thus, while African cities have traditionally been associated with "informal sector" activity, they are now experiencing new waves of informality (Lindell 2010; Meagher 2010). The poor are devising creative ways of coping with the economic crisis.

In the context of informal solid waste (ISW) recycling, there is a need to understand how operators organise and try to maximise their opportunities. This chapter attempts to analyse the social process of ISW recycling in Zaria, Kaduna state, Nigeria. It aims to show the broader issues of informality, often disregarded, involving a variety of relations and linkages. Furthermore, it sheds light on how the various actors are faring and their survival strategies. The analysis is based on a review of literature on informal waste recycling and the author's field work in 2012 (Ukoje 2012) involving a survey of 320 ISW operators to gather information on their organisation, relationships and strategies for survival.

Informal Solid Waste Management: An overview

ISW management is prevalent in African, Asian and Latin American cities (Medina 2007). In this context, Wilson et al. (2006) describe the ISW sector as the waste recycling activities of waste pickers and waste processors. These terms refer to those involved in the extraction of recyclable and reusable materials from mixed wastes, storage in junkshops and processing (Bernstein 2004). Such activities epitomise the informal sector as they are labour intensive, low technology, low paid, unrecorded and unregulated, with the work often being undertaken by individuals or family groups (Wilson et al. 2006). Scheinberg (2001) observed that more complex organisations can develop, such as micro and small enterprises involving groups of up to 10 (micro) or 20 (small) people, and operating with low capital investment and technology. Consequently, entry into the trade is easy. This means that even those unable to work full time, such as children and women, or those formally employed but having some spare time or requiring extra income, can also engage in ISW income-generating activities.

The above account highlights how ISW operations are an ideal strategy for making a living and for helping to achieve one of the aims of modern waste management, namely moving up the waste hierarchy by reducing reliance on disposal through recycling (Wilson et al. 2006; Medina 2000). Because the poor and unemployed lack basic assets and capabilities to earn a living, waste materials become resources that can be used to generate an income flow or other benefits to the economy.

Previous studies tended to emphasise the low income derivable from informal sector activities (Birkbeck 1978, 1979; Tevera 1994). However, recent research shows that ISW activities have become economically viable ventures (ILO 2004; Wilson et al. 2006). Several authors in Nigeria stress their contribution to poverty reduction, the achievement of socially related Millennium Development Goals and livelihood diversification (Agunwamba 2003; Nzeadibe 2009; Ukoje 2012). Unfortunately, researchers have devoted little attention to the complexity and dynamism of the associated social relations and power networks (Adama 2012). This has necessitated a study of the social processes operating in the sector. Understanding these processes will direct policy towards the sector with the aim of significantly enhancing its performance.

Zaria and Informal Solid Waste Recycling

Zaria is home to the famous Ahmadu Bello University and to a host of federal institutions, and has grown rapidly over the decades in population, economically and in spatial extent, largely as a result of the numerous socioeconomic and educational opportunities it offers. Its population, according to the 2006 census, is 975,153, spread across two local government areas, Zaria and Sabon Gari. There are four distinct urban districts: Zaria City, Tudun Wada, Sabon Gari and Samaru. Zaria City retains the physical character of a traditional indigenous settlement of Muslim Hausa-Fulani form (Urquhart 1970). Tudun Wada is inhabited by a mixed indigenous and non-indigenous population, while Sabon Gari, the functional commercial centre of Zaria, is largely populated by migrants from the south of the country. Samaru has a mixed population of indigenous Hausa and migrants and is home to Ahmadu Bello University, the largest university in northern Nigeria.

In Zaria, ISW trade and recycling are largely linked to the fortunes of the people working or studying in the area. The quest for employment has spurred the influx of migrants from other parts of Nigeria and West Africa to the town. However, while the city's educational institutions facilitated urbanisation there are fewer non-professional job opportunities for less educated indigenes and migrants, who constitute the majority of the population. In response to the difficulties of securing formal sector employment, people resorted to informal services, including ISW recycling.

Structure of Informal Solid Waste Recycling in Zaria

ISW in Zaria covers a wide range of activities, including waste collection, waste picking on streets and dump sites, itinerant waste buyers, sorting, storage and compacting by small and larger junkshops and actual reuse and recycling among practitioners in their drive for survival. Across this range, a significant number of children and men participate, not only for the survival of their households but also, perhaps even more so, as the only means of employment. Based on the findings of this study, limited access to operational resources has tended to concentrate participants at the low income end of the waste trade. In order to increase household income and improve family well-being, operators resort to other measures for survival.

Types and Functions of ISW Activity in Zaria

A complex chain of ISW stakeholders works on a small scale in collecting, storing, recycling and processing waste for wealth creation and environmental protection. These activities fall into three main categories: waste picking; waste storage/junkshops/middlemen; and reuse and recycling (Table 6.1). In several cases, there is more than one type of activity in one category. Most individuals belong to the waste picking category and work as itinerant waste buyers and collectors, waste pickers on streets and dumps and as direct collectors of waste. This category of ISW activity is associated with selective waste collection, including recovery and separation of recyclable materials from mixed waste, an indirect source of secondary raw materials for local industries.

Table 6.1:
ISW activities and functions

Activity	Functions
Waste picking (scavenging)	Recovery of materials Low cost collection Provision of secondary raw materials
Junk shops/middlemen	Waste storage Sorting and separation Compacting/reduction in volume Intermediaries between waste pickers and recycling industries
Reuse and recycling industry	Utilise recovered materials as feedstock Provide cheap products

Source: Field work 2012

Junkshops/middlemen are fewer than the first category, and act as dealers for the collected materials. They operate small- and large-scale junkshops and are associated with the sorting, temporary storage and compacting of select waste materials. Data for this study indicate some specialisation in the items collected by the junkshop owners. While some junkshops specialise in plastics, bottles and scraps, most deal in all items. The dealers sort and separate waste into its component parts, thereby adding value to the waste. Some dealers compact and physically reshape the products to reduce their volume for ease of handling and transport. The junkshop/middleman category serves as intermediaries between waste pickers and the reuse and recycling industries.

Reuse and recycling constitutes the last and least numerous categories. It comprises aluminium foundries, metal fabricators and some formal private enterprises that produce plastics, tissue paper, etc. These industries use the materials provided by the first two categories for reuse, trade and as feedstock for manufacturing. Some of their recycled products, such as pots and buckets (Figure 6.1) are cheaper than the equivalent products manufactured in modern industries.

Generally, ISW recycling activities demonstrate the economic benefits of the provision of secondary raw materials to industry and are clearly important as a form of livelihood for so many households. In addition, the sector performs an important role in reducing the cost of waste disposal and the environmental problems associated with uncollected waste.

Figure 6.1:
Recycling aluminium at one of many centres in Zaria. (Author's photo)

Social and Human Dimension of ISW Recycling

ISW recycling is undertaken by both young and old, but is dominated by men. This study revealed weak representation of women, who account for only 14 per cent of those in the waste picking category. Most of these are girls and the elderly (Table 6.2). Male pickers account for 86 per cent of the category. This contrasts with the findings of Agunwanba (2003) and Nzeadibe (2009) who, using a sample of 13 and 24 interviews respectively, reported that 77 per cent and 67 per cent of waste pickers in their study areas were men. The relative ease of entry is an important factor for the predominance of children and those lacking in investment capital. However, entry into higher-level activities such as storage, middlemen and recycling seems to be more difficult, owing to the capital costs, and only adults are involved. In fact, men constitute 95 per cent of those engaged in intermediate categories and 98 per cent of those engaged in reuse and recycling.

Table 6.2:
Socioeconomic characteristics of participants in ISW activities in Zaria

Aspects	Characteristics	Percentage of Activities		
		Waste Picking	Inter-mediate Dealers	Reuse and Re-cycling
Gender	Female Male	14 86	5 95	2 98
Age	< 18 years > 18 years	68 32	- 100	- 100
Education-al Level	Primary Secondary Tertiary Not Literate	7 - - 93	28 6 - 66	36 13 - 51
Average Monthly Income in Naira(₦)*	< 7,000 7,001-11,000 11,001-15,000 15,001-20,000 > 20,000	100 - - - -	- 6 13 30 51	- 7 18 33 42

Source: Field Work 2012

*At the time of writing, US$ = ₦150 at the official rate.

The level of formal education of operators is very low, particularly the waste pickers, only 7 per cent of whom have primary education. The most educated are the reuse and recycling operators, of whom 36 per cent have primary and 13 per cent secondary education. Interestingly, ISW activities do not require much formal education (Wilson et al. 2006), except those related to recycling. Even at this end of the spectrum, all the operators learned their skills on the job, as all the fabricators, foundry workers, etc. were once apprenticed in the activities they now undertake. This result does not suggest that formal training is useless, but rather indicates its limited usability outside the formal sector, and the necessity for informal retraining for operators in the sector.

Those working in the informal sector have generally been considered to earn a marginal income compared to those in the formal sector (Meagher and Yunusa 1996; Ahonsi 2000). While this may be true of waste pickers, who earn less than ₦7,000 per month, 81 per cent of intermediate dealers and 75 per cent of recyclers earn incomes that compare favourably with formal sector wages. For instance, the minimum monthly income in the civil service is ₦18,000. This suggests that ISW is a profitable occupation in Zaria and, since daily earnings exceed US$1, operators may be taken to be above poverty level.

ISW Recycling Organisation and Survival Strategies

The ways in which ISW operations are organised has important implications for income generation and survival. Waste pickers and scavengers are easily exploited by intermediate dealers and the vulnerable are mainly children, women and the elderly. About 66 per cent are actively recruited by dealers, who offer loans to the recruits. However, those who stand on their own sell directly to end users and earn double the profit of those who are recruited. The survey elicited data on the means of generating cash incomes. Taking on more than one income activity, though not restricted to the informal sector, is central to survival in the sector. Despite the long hours of work in ISW recycling, many waste pickers and dealers engage in other activities to earn additional income. Of the sampled operators,

71 per cent pursued at least one other activity. Farming is the most significant such activity, and is practised by 69 per cent of those who undertake other activities. The percentages for petty trading and sewing are 21 per cent and 8 per cent respectively. Less common other activities include brick making, dressmaking, car repairs and poultry raising. Those at the top of the secondary material trade hierarchy are less likely to engage in other income-earning activities: over 83 per cent of them do not, as they reported an adequate income from their main activities.

The middlemen and craftsmen occupy an intermediate position in the hierarchy and serve as a link between waste pickers and end users. This category includes dealers, recycling operators, junkshop owners, intermediate processors and wholesalers, all of whom undertake a range of value-added activities such as sorting, cleaning, pre-processing, and reshaping materials to facilitate transport and storage so that commercially viable volumes of raw material can be aggregated. Through these activities, intermediate operators increase the value of the secondary raw materials they trade in and earn higher incomes. In addition, many dealers are able to generate a regular income through their established linkages with the formal manufacturing industries, which depend on the secondary raw materials the sector supplies. These companies give credit in advance to the dealers to ensure a stable supply. One good example is the toilet tissue manufacturing company in Aba, which provides regular credit advances to a major dealer in Zaria to enable him to buy up old newspapers and used papers from individuals and academic institutions in the town.

Participation in informal unions is one of the coping strategies adopted by informal sector workers (Ahonsi 2002: 145; Meagher and Yunusa 1996) and enables members to adapt to social and institutional constraints (Meagher and Yunusa 1996). Unionisation appears significantly more important among junkshop owners and recyclers than waste pickers. Throughout Zaria, none of the waste pickers sampled belonged to a union, while 84 per cent of the junkshop owners and 91 per cent of recyclers participate in collective organising activities. Through these social networks, recyclers are able to buy raw materials in bulk at cheaper rates than at an individual level, resolve conflicts among themselves, maintain peace and protect members against crime and theft. In Zaria, the union of recyclers formed the avenue along which the Kaduna state government under former Governor Makarfi advanced financial aid to its members. This union gave advance loans to about 21 per cent of its members at various times to upgrade their activities. Some of the older recyclers and their apprentices have expressed willingness to work through their association to establish a waste recycling plant so as to reduce their dependence on middlemen and therefore increase their profit margins.

At the top of the waste trade are the manufacturing industries that are strongly dependent on secondary raw materials. These are mainly formal industries, which in response to local sourcing drives and the high cost of imported inputs, are forced to use recycled materials collected by the informal sector. These small-scale industries deal directly with middlemen and promote and finance the recovery and recycling of materials.

Challenges to Informal Solid Waste Recycling Operators

Studies by Haan et al. (1998), Scheinberg (2001: 525), ILO (2004: 20) and Bernstein (2004: 126) indicate the problems ISW operators encounter in their operations. The present study seems to support these conclusions, as operators identified a number of difficulties they struggle within their businesses.

Limited Credit and Capital

The cost of starting up an ISW operation is high, particularly an intermediate junkshop and recycling plant. While waste picking requires very little capital, recyclers and dealers need between ₦50,000 and ₦100,000 for basic equipment and to rent premises. The interviews show that these start-up costs are beyond the means of most of the respondents. Initial capital outlay was obtained from personal savings, friends or relatives. In the study, 63 per cent of operators drew their start-up capital from their own personal savings, 31 per cent obtained it from relatives and 6 per cent from friends. None of the operators got credit facilities from financial institutions. As a result, most operators resort to old and locally fabricated equipment that needs constant repair.

Limited start-up capital and credit facilities explain why these activities are small scale and are undertaken in appropriated open spaces and dilapidated buildings. In fact, the operators are unwilling to take out loans from financial institutions because of the attached conditionalities. This suggests that any credit support is obtained through other channels.

Because of their limited financial resources, ISW industries buy their inputs locally and in small quantities. Raw materials in Kaduna and Abuja are relatively lower in price when purchased in bulk. For instance, while scrap aluminium costs ₦150 per foot locally, bulk purchases from Kaduna, Abuja and southern Nigeria are 25 per cent cheaper. Respondents identify input prices as a significant part of the cost of production, while lack of access to credit facilities hinders them in cutting down these costs and reduces their profit margins.

Inadequate Physical Infrastructure

High rents and expensive urban utilities are frequently cited as problems for informal activities. Local government taxation of businesses and revenue-collection drives are a worry for operators. Tax rates are being raised to increase revenue, but government has become less able to maintain or expand physical infrastructure, such as water and roads. Respondents operating junkshops complain of erratic water supplies for their cleaning operations. They have to pay water vendors to bring water to them from commercial wells and boreholes. In addition, electricity bills and rents have risen and for most operators landlords now demand 12-month advance payments. To cope, operators resort to squatting on open spaces within self-erected and ramshackle structures.

Government and Public Attitudes

The bulky nature of waste materials requires accessible locations along roads. However, local government and town planning officials do not officially recognise the operators and are more concerned with the aesthetic environment. This tends to create tension among the two groups. For instance, 65 per cent of junkshop owners have complained of harassment by government officials, who forced them to relocate to inaccessible locations where it is difficult to load and off-load their goods. In addition, public attitudes toward waste pickers and waste recyclers have been negative. They are looked down upon and sometimes wrongly accused of stealing as they go about their trade.

Conclusion

This study of the ISW recycling in Zaria has highlighted several processes at work in the pursuit of livelihoods in the sector. The generally low levels of urban employment force a large number of people to rely on informal income generation. Coincidentally, the ready availability of solid waste associated with insufficient collection and disposal services presents an opportunity for the unemployed to earn an income by recycling and providing secondary raw materials. In addition, the strong dependency on secondary raw materials by certain formal manufacturing industries has expanded survival opportunities for the operators. It can therefore be argued that ISW is making an important economic contribution in Zaria by creating employment and reducing poverty. However, the obvious differences across gender, age, type of activity and level of income generation call for different approaches to the needs of the various groups.

In addition, this study has shown both elements of indifference towards and symbiotic relationships with ISW operators. While the public sector seems not to be concerned about the plight of the sector, the private sector is positively disposed towards the informal sector. Research is needed to identify approaches that the public sector can adopt to assist ISW operators and mitigate the challenges facing them. Their economic contribution to Zaria could be significant if the social processes are improved. Supportive policies need to be developed to stimulate and improve relationships with and working conditions within the sector. Local, state and national governments have key roles to play in collaboration with non-governmental agencies, community-based organisations and, on occasion, international donor agencies.

References

Adama, O.A. (2012). Urban livelihood and social networks: Emerging relations in informal recycling in Kaduna, Nigeria. *Urban Forum* **23** (4): 444 – 66.

Agunwamba, J.C. (2003). Analysis of scavengers' activities and recycling in some cities of Nigeria. *Journal of Environmental Management* **32** (1): 116 – 27.

Ahonsi, B.A. (2002). 'Popular shaping of metropolitan forms and processes in Nigeria; Glimpses and Interpretations from an Informed Lagosian', In *Under Siege: Four African Cities, Freetown, Johannesburg, Kinshasa, Lagos, March 16-20, 2002.*, O. Enwezor, C. Basualdo, U. M. Bauer, S. Ghez, S. Maharaj, M. Nash and O. Zaya (eds). Documenta 11, Platform 4. pp. 129 – 52.

Bernstein, J. (2004). *Social Assessment and Public Participation in Municipal Solid Waste Management: Toolkit.* Washington DC: Urban Environment Thematic Group, World Bank and National Academy Press.

Birkbeck, C. (1978). Self-employed proletarians in an informal factory: The case of Cali's garbage dump. *World Development* **6**(9-10): 1173 – 85.

Birkbeck, C. (1979). Garbage Industry and the "Vultures" of Cali, Columbia. In Casual Work and Poverty in Third World Cities, R. Bromley and C. Gerry (eds.). New York: John Wiley, pp. 161 – 83

De Soto, H. (1989). *The Other Path: The invisible revolution in the Third World*. New York: Harper and Row.

Haan, H.C., A. Goad and I. Lardinois (1998).*Municipal waste management: Involving micro-and small enterprises. Guidelines for municipal managers.* Turin: International Training Centre of the ILO, SKAT, WASTE http://www.skat-foundation.org/publications/waste.htm.

ILO (2004).*Addressing the Exploitation of Children in Scavenging (Waste Picking): A Thematic Evaluation on Child Labour*. Geneva : International Labour Office.

Lindell, I. (ed.) (2010). *Africa's Informal Workers: Collective agency, alliances and transnational organising in urban Africa*. London/Uppsala: Zed Books and Nordic Africa Institute, pp. 1 – 30.

Meagher, K. (2010). *Identity economics: Social networks and the informal economy in Nigeria*. New York: James Currey.

Meagher, K. and M.B. Yunusa (1996). Passing the Buck: Structural Adjustment and Nigerian Urban Informal Sector. Discussion Paper No.75. Geneva: United Nations Research Institute for Social Development.

Medina, M. (2000).Scavengers' cooperatives in Asia and Latin America. *Resources and Recycling* **31**(1): 51 – 69.

Medina, M. (2007).*The world's scavengers: Salvaging for sustainable consumption and production*. Lanham MD: AltaMira Press.

Nzeadibe, T.C. (2009). Development drivers of waste recycling in Nsukka Urban Area, Southeastern Nigeria. Theoretical and Empirical Researches in Urban Management **3** (12):137 – 49.

Scheinberg, A. (2001). *Micro-and small enterprises in integrated sustainable waste management.* The Netherlands: WASTE. http://www.waste.nl/page/534.

Sethuraman, S.W. (ed.), (1981). *The Urban Informal Sector in Developing Countries*. Geneva: ILO.

Tevera, D.S. (1994). Dump scavenging in Gaborone, Botswana: Anachronism or refuge occupation of the poor? *Geografiska Annaler. Series B, Human Geography* **76**(1): 21 – 32.

Ukoje, J.E. (2012). Informal sector solid waste collection and recycling in Zaria, Nigeria.*Journal of Environmental Science and Engineering* **B 1**: 649 – 55.

Urquhart, A.W. (1970). Morphology of Zaria. In *Zaria and Its Region*, M.J. Mortimore (ed.). Occasional Paper No. 4, Department of Geography, Ahmadu Bello University.

Wilson, D.C., C. Velis and C. Cheeseman (2006). Role of informal sector recycling in waste management in developing countries. *Habitat International* **30**(4): 797 – 808.

7

PROSPECTS AND RETROSPECT: INSIGHTS FROM CHILDREN'S PARTICIPATION IN INFORMAL SECTOR SOLID WASTE MANAGEMENT IN KADUNA, NIGERIA

Adamu Ahmed, Onyanta Adama and Idris Isa Aliyu

Introduction

Children and youths constitute a large part of the global urban population. In Africa, Asia and Latin America, over 80 per cent of the population is accounted for by children (UNICEF 2002). In their urban domain, children and youths are challenged by numerous difficulties, which usually influence the way they behave. Such behaviours have been the subject of debate, ranging from issues of negative behaviour, rights and the kind of work children should partake.

There is a large literature documenting child involvement in drugs and violence (especially when children lack opportunities), and also of their productive contributions (Nieuwenhuys 1996;

Potts 1997). Boyden (2003) associates negative behaviour of children with the lack of protection of rights and access to basic means of livelihood, about which several pieces of policy and legislation exist. Bourdillon et al. (2010) observe that children attempt to negotiate livelihoods only when their survival needs are denied.

Perhaps a much bigger debate has been about the work children do to survive and how such work can be situated within rights charters (Thorsen 2012). This concern follows from observations that some forms of child work can be beneficial to children's well being, while others are detrimental or hazardous. Children are acknowledged to have few resources, material or social, with which to meet the threat of poverty in African cities, and are therefore particularly likely to suffer in crises arising in those cities. However, cities also provide opportunities for them (Honwana and DeBoeck 2005). In spite of several constraints, many young people find livelihoods in cities, which they prefer to what is available in rural areas. That is, rather than submitting passively to poverty they frequently find employment and are often a major resource in dealing with urban problems and in developing the cities. In this chapter, further insight is provided into the complex forces children contend with in negotiating their survival in cities through waste picking.

Children in the informal economy

The role of children in the informal economy continues to generate much debate, largely due to the issue of child labour. International Labour Organisation (ILO) Convention 182, which aims to eliminate child labour, lists the worst types of such labour as all forms of engagements or practices similar to slavery; the use, procuring or offering of a child for prostitution; the use, procuring or offering of a child for illicit activities; and work which, by its nature or other circumstances, is likely to harm the health or safety of children. The number of working children aged between 5 and 14 years in the world is about 211 million (ILO 2002). Of this number, sub-Saharan Africa has the highest proportion, about 29 per cent. The ILO also reports that children are often involved in hazardous and exploitative work. Cities are a focal point and the informal sector is an important arena of employment for children (Thorsen 2012). Children are involved in a range of activities, including waste picking, family or home-based work, manufacturing and hard physical labour, such as construction (Holly Te Consultant 2005). They tend to work alongside a parent, have little social protection, are subject to abuse and violence and some work without pay (Medina 2011). There is also a gender dimension. Of the 211 million child workers given, about 102 million are girls (ILO 2002). The minor variation in gender composition is reflected in Phnom Penh, where the ratio is 53 per cent girls to 47 per cent boys (Holly Te Consultant 2005). However, there are marked gender differences in relation to type of activity. In Uganda, more girls than boys work in bars, restaurants and in domestic service, while there are more boys in hawking, carpentry and waste picking (ILO 2004). More importantly, girls tend to be more vulnerable than boys. Girls start work at an earlier age, are paid less and are particularly exposed to exploitation and abuse, including prostitution (ILO 2002).

The reasons children work in the informal economy have received considerable attention in the literature. Poverty is cited as the most important reason among children of school-going age (Thorsen 2012). This argument is supported by the contention that child labour tends to increase during economic crises, as more and more children enter the informal economy to earn money and contribute to the family income (ILO 2002). However, poverty is not the only factor. Some children join the informal economy for income to pursue formal education or to learn a particular skill, and do not necessarily have to drop out of school (Thorsen 2012). Some combine part-time work and education. Overall, education plays a positive role, since the greater the number of children enrolling in schools, the lesser the number engaging in informal activities (Meagher and Yunusa 1996). The high percentage of children in the informal economy is also linked to the notion that children offer cheap labour to men and women. This is particularly relevant in places where women's mobility is constrained, for example, in northern Nigeria, where the majority of Muslim women have limited access to urban space (Thorsen 2012). In such cases, children are useful because they can enter the gendered spaces of both men and women. Another avenue through which children enter the informal economy and become sources of cheap labour is the outsourcing of production by big firms to developing countries. The pressure of global competition and the need to reduce costs is a major explanation for transnational corporations outsourcing component production to different locations in developing countries. Garment and shoe factories in Asia are noted for their deplorable working conditions.

The determination of safe and unsafe work that children should do has been the subject of much recent debate. Certain forms of child work are known to be more beneficial than detrimental to children's well being, just as some work may be hazardous and strenuous. In a related development, the concept of decent work has been popularised by the ILO. The term is often used to highlight the plight of both formal and informal workers, but the emphasis is on the latter. In the case of the informal sector, the ILO argues that the most meaningful approach is to talk about work deficits, such as poor quality and unremunerated jobs that are not protected by law; the absence of rights at work; inadequate social protection; and the lack of representation and voice (ILO 2002). Crucially, the ILO adds that the problem is most pronounced among women and children. Children and youths form the majority of waste pickers in cities across the globe (Medina 2011), which makes it necessary to raise the issues of child rights and decent work in the Informal Waste Management (IWM) sector.

Study Context

Kaduna metropolis is a cosmopolitan City of about 1.6 million people. It was previously the colonial capital of the northern Nigeria protectorate. Since independence, it has slipped from being a regional capital of North Central State to its current status as capital of a much smaller territory, Kaduna state. Kaduna, with its cluster of industries and its geographical location, has also been of economic significance. Typical of settlements of its status, Kaduna has a huge migrant population, a sizeable part being youths living under parental care or as itinerant labour. Among these youths are the Almajiris, young Muslim pupils in the Islamic education system. Abandoned by their parents and largely ignored by the government, they are among the most disadvantaged and vulnerable groups in the city. Their growing number and the ways they negotiate their livelihoods have been a source of concern, as have the new forms and styles of urban life and sociology they have influenced.

Torn between illusion and reality, children across most urban areas, including Kaduna, have been compelled to negotiate livelihoods under sometimes extreme circumstances. In the case of waste picking, failure to wear protective clothing is widespread among all age groups, but children are particularly vulnerable (Adama 2012; Holly Te Consultant 2005). For example, they have thinner skin layers and a faster rate of breathing and are more vulnerable to airborne disease (Hunt 1996). Added to these is their relatively young age and lack of judgement, knowledge and experience, which often leads them to take impetuous decisions and hazardous actions.

Children's efforts to survive through activities in the subsistence sphere of the peasant economy are widely acknowledged in the literature (Nieuwenhuys 1996; Boyden 2003). This includes studies of their productive capacity to help themselves and their households. In traditional African societies, for instance, children are required to contribute to the family's means of subsistence by looking after the home and contributing their labour in agriculture and animal husbandry. Furthermore, as reported in the case of Nairobi's Mathare Valley, waste is not only a major source of income generation for youths, but also a means of asserting their rights to the city (Thieme 2010).

The study reported in this chapter contributes to the debate on child labour in general and the issues raised above in particular. It seeks to establish how children negotiate livelihoods through waste collection and picking, the challenges they face, and the health, environmental and urban development implications of their involvement. The study adopted the mixed methods approach. It relied on field visits, use of questionnaires, and conduct of interviews with children, vendors, households and municipal authorities. A total of 10 dump sites were visited and 40 questionnaires were administered on location.

Children in the informal waste economy of Kaduna

In Kaduna, and in line with the global norm, children and youths form the largest proportion of waste pickers. Children engaged in waste picking are between the ages of 8 and 19 and are predominantly male. Most of the children (65 per cent) got into waste picking through peer influence for survival. Parental influence was in play for 35 per cent of the children. The majority of them (58 per cent) in waste picking are migrant youths. About 38 per cent are Almajiris. The activities of child waste pickers are not formally organised, although networks are maintained with many vendors that handle waste recycling. Children participate at two levels: house-to-house collection and at the dump site. A third group participates at both levels. House-to-house collection is a major activity for about 32 per cent of the sampled children. Participants, typically called "barrow boys," (see Fig 7.1) empty the waste bins of households for a token and according to agreed routines. The equipment used is limited to a wheelbarrow and nothing more, which often (65 per cent of cases) is rented.

Figure 7.1:
Barrow boy at a collection point at Ungwan Rimi (Authors' photo 2011).

Children present at dumpsites as waste pickers represent 82 per cent of the total sample. Unlike house-to-house collectors, they lack any sort of equipment, except the polythene bags they carry about for storage. The primary clientele of this group is the middlemen and vendors, who usually purchase recovered valuables onsite. The third group (35 per cent) intermingle house-to-house collection with waste picking at dumpsites. Studies show that across the globe, child waste pickers work long hours (Medina 2011; Hunt 1996). In Kaduna, time spent on waste picking varies between two and four hours daily, usually in the mornings (7:00 – 11:00am). In this period, activities are sequenced: house-to-house collection is undertaken in the early hours and sorting at dumpsites later. The former peaks between 7:00 am and 9:00 am. This is when households do their cleaning. Waste picking occurs between 9:00 am and 11:00 am, when collections have reached the sites. During the peak hours of waste picking, adults are the primary participants and children are usually allowed to participate only after adults have vacated the site.

Benefits of Children's participation in Waste Management

Personal

Waste picking has provided social and economic advantages to most of the children and their parents. The average income may not be significant, but it usually takes care of some of the needs of children, normally a parental responsibility. About 55 per cent of children claim to be making remittances or contributions to their parents. Income varies between house-to-house collection and waste picking. Higher returns accrue to the former, between ₦ (Naira) 200 (the equivalent of US$ 1.26) and ₦ 250 (US$ 1.57) per day compared to ₦ 150 – 200 (US$ 0.94 – 1.26) from waste picking at the dumps (US$ equivalents calculated on an exchange rate of ₦1 to US$0.0062 obtained on November 11 2013). Involvement in waste picking is not a primary vocation for most of the children, but supplements other remunerative activities they engage in. These include house cleaning, errands, hawking and other menial jobs. The limited participation of this group of children in waste picking as opposed to waste collection is linked to the power dynamics among the children. While children as a group face harassment from adults, the youngest children are often subjected to intimidation by their older peers. The youngest children are seen as a threat especially in relation to the recovery of the more valuable materials. It is therefore common for the older ones to adopt all kinds of strategies to keep them away. The value of waste collection and picking for children appears obvious in that it has helped them to become productive, resourceful and disciplined. Many children have also used waste picking especially for socialising, by mixing work and play. At most dump sites, children could be seen playing games for an unusually long time to make up for the lack of formal recreational opportunities and to cope with unsatisfactory work situations. This is valuable for self-development.

Societal

Although precise estimates are difficult, the amount of waste handled by children appears to be quite appreciable. Primarily through house collections, children are apparently making a huge contribution to the management of wastes in the metropolis. The average weight of waste handled by barrow boys per household serviced was estimated at between 2 and 3 kilograms in the Unguwar Rimi neighbourhood. In many parts of the city not covered by private collectors, the "yan barrow" are the army of collectors keeping households clean of wastes. Interestingly, compared to private, formal waste collection companies, the study shows the "yan barrow" to be more reliable and cheaper. Waste picking by children also has utilitarian value at the town level through the discipline, social harmony and resourcefulness it helps foster. As a result of such productive engagement, there have been fewer incentives for children to engage in crime, drugs and social vice.

On the material side, the study corroborates existing literature (Zelizer 2002) showing children to be economically productive on the basis of the work they undertake. In traditional rural African societies, children have long been a major economic force through their contribution to family subsistence. This contribution has ranged from looking after the home and young children, to contributing labour to agriculture and animal husbandry. In many urban areas, children also contribute to family incomes in many of ways. Indeed, the African Charter on the Rights and Welfare of the Child points to the duty of children "to work for the cohesion of the family, to respect his parents and elders at all times, and to assist them in case of need" (OAU 1990).

Hazards of Waste Picking

Waste picking involves hazards that the children are generally unaware of. Only 15 per cent acknowledge concerns about safety and hygiene. Protective clothing is unheard of and waste collection and scavenging is undertaken barehanded. About 38 per cent of the children acknowledge they have been injured while handling waste. Coughs, skin irritation and eye infections are common among the children, but they do not relate it to waste handling. The study could not reach the parents of the children interviewed to establish their understanding of the risks associated with their children's vocation. However, since parents are mostly poor, the pressure for survival would most likely override any safety concerns for their children.

Other than the health risks, the effects of child work on education have potential long term consequences for society. Long hours spent by youths outside school means that the future productive labour requirements of the city will be jeopardised. Cities are likely to be trapped in a cycle of poverty with limited capacity to guarantee better livelihoods. Also apparent from the study are the environmental hazards at the town level arising from children's informal solid waste management activities. Limited in reasoning capacity, many children usually dump collected refuse in open spaces, streams or along road reservations. Few (11 per cent) would own up to this, but the practice appears widespread. Equally, the burning of refuse by children has environmental implications.

Waste Picking and Contesting Urban Space

Children, like adults, are generally not restricted as to where they pick or collect waste. That is, spatially controlled territories do not exist, but each player usually survives by negotiating his way. With house collections, the ability to retain clientele by providing decent services acts as a form of control. Dumpsites are no-man's land, but adults frequently harass younger participants. In respect of house collections, the study shows that children usually compete not only with adults, but with private-sector commissioned collectors. Although not necessarily consciously so, children are able to negotiate participation by being cheaper and by providing more reliable services. By charging between ₦ 20-50 (US$ 0.12 – 0.31) per collection, children are able to deliver services at a fraction of the charges levied by private companies. As noted earlier, children are able to operate at dump sites only after the adults have left, usually between 10.00 am and 1.00 am, by which time most of the valuables have been carted away. The only rule operating in this regard is the social construct defined by adult worldviews that children's work is deemed inferior to adult work.

A third level of contestation confronting children relates to price negotiations. House collection fees charged by children are usually lower than those sought by adults, especially as at times children may view food leftovers put out by families as part payment. At dumpsites, children sell recovered items mainly to middlemen, who capitalise on their innocence by paying less. The fourth level of contestation is with municipal authorities, who usually chase children away from dump sites for what they claim is unlawful behaviour. This relates to the burning of refuse, a practice the authorities have banned. Children and youths are often in a disadvantageous position when bargaining with municipal authorities, as has been noted with respect to the youth garbage collectors of Nairobi, Kenya (Thieme 2013). The youth there have little faith in official political engagement as a means to address their issues. Their inability to form a broad network has resulted in an uncoordinated mosaic of small-scale enterprises unable to access capital, effectively engage with broader issues or defend members against harassment by municipal authorities. In Kaduna also, children are not formally organised and therefore lack a platform to articulate their needs and protect their interests.

Conclusion: Rethinking Children's Participation in Waste Picking

The poverty of families across most of Africa has compelled children to work in various ways to supplement livelihoods for themselves and their households. This situation, although not consistent with various charters of the rights of children, has been explained on the grounds of the many shortcomings of society and governance. Many NGOs committed as a result of this to protecting children's rights have been criticised for being too dogmatic and inflexible in their approach. It is often held that they do not always work in the best interests of children. In many parts of Africa, as Jacquemin (2006) observes, commitment by organisations to a particular interpretation of child rights has worked against the children most in need of support. On the basis of this argument, rethinking children's work is here advocated: the general problems international conventions seek to address do not always reflect the thinking, values and circumstances of particular communities or of the children they are intended to protect. This issue is reflected in the African Charter, which links children's rights to children's responsibilities towards their families, communities and their societies (OAU 1990). The Charter speaks of taking due account of the traditions and values of peoples in seeking "the protection and harmonious development of the child" (OAU 1990).

Next, not all work is actually work for children. As this study shows, waste collection and picking does not necessarily fall into the category of work, despite obvious challenges and risks. It has rather been beneficial in allowing children to meet basic needs and support families and by inculcating in children a higher sense of responsibility, self-worth and a sense of satisfaction. However, there are risks that require careful consideration. First are the health risks. Unaware of the wider implications, most of the children approach waste picking without adequate care. They handle wastes with their bare hands and without protective clothing. This study has not proven that their health problems are associated with waste picking, but the probability is that there is a close relationship. Secondly, with children staying out of school, the city's future is placed in jeopardy by an army of uneducated and unproductive labour. As observed by Bourdillon (2009), the future of African cities depends on their large youthful populations. They will determine what cities become by who they are.

In dealing appropriately with child work in the form of waste picking, society will need to strike a good balance between the challenges and potentials. A sound answer will depend on good urban governance, legislation, partnerships and innovative strategies. Children's waste-picking activities to support themselves and their families will especially require to be reconciled with their need to remain in school. That way, better participatory outcomes can be achieved.

References

Adama, O. (2012). Urban Livelihoods and Social Networks: emerging relations in informal sector recycling in Kaduna, Nigeria. *Urban Forum* **23**(4): 449-466.

Bourdillon, M. (2009). A place for work in children's lives? PLAN-canada. Available at http:// plancanada.co

Bourdillon, M., D. Levison, W. Myers and B. White (2010). *Rights and wrongs of children's work.* New Brunswick: Rutgers University Press.

Bourdillon M. and P. Hebinck (2002). The concept of Livelihood. In *Women, men and work: Rural livelihoods in South eastern Zimbabwe*, M. Bourdillon and P. Hebinck (eds). Harare: Weaver Press.

Boyden, J. (2003). Children under fire: Challenging Assumptions about children's resilience. *Children, Youth and Environment* **13**(1): 1-29

Holly Te Consultant (2005). How and Why We Work: Child Workers in the Informal Economy in Phnom Penh and Battambang. Combating the Worst Forms of Child Labor Project Peace and Justice Programme.World Vision Cambodia http://www.worldvision.org.kh/pdf/Child_Workers_in-the-Informal_Economy.pdf .Retrieved 20 March 2014.

Honwana, A. and F. Deboeck (2005).*Makers and Breakers: Children and Youth in Post Colonial Africa.* Oxford and Dakar: James Currey and CODESRIA.

Hunt, C. (1996). Child waste pickers in India: The occupation and its health risks. *Environment and Urbanization***8**: 111 – 18.

ILO (2002). *Decent work and the informal economy.* Geneva: ILO Publications.

ILO (2004).*Report of the Sectoral Study on Child Labour and the Urban Informal Sector in Uganda.* Geneva: International Labour Organisation.

Jacquemin, M. (2006). Can the language of rights get hold of the complex realities of child domestic work? The case of young domestic workers in Abidjan, Ivory Coast. *Childhood* **13**(3): 389-406

Meagher, K. and M. Yunusa (1996).*Passing the Buck: Structural Adjustment and the Nigerian Urban Informal Economy.* UNRISD Discussion Paper 75. Geneva: United Nations Research Institute for Social Development

Medina, M. (2011), The informal sector – a driving force for recycling management. In *Recovering resources, creating opportunities: integrating the informal sector into solid waste management.* E. Gunsilius, S. Spies and S. García-Cortés (eds.). http://wiego.org/publications/recovering. Retrieved 18 December 2011.

Nieuwenhuys, O. (1996). The paradox of child labour and anthropology. *Annual Review of Anthropology* **25**: 237-251

OAU.(1990). African Charter on the Rights and Welfare of the Child.OAU Doc. CAB/LEG/24.9/49 (1990).

Potts, D. (1997). Urban lives: Adopting new strategies and adapting rural links; in *The urban challenge in Africa: Growth and Management of its Large Cities.* Carole Rakodi (ed.). Tokyo: UN University Press.

Thieme, T. (2010). Youth, waste and work in Mathare: Whose business and whose politics? *Environment and Urbanization* **22**: 333 – 52.

Thieme, T. (2013).The "hustle" amongst youth entrepreneurs in Mathare's informal waste economy. *Journal of Eastern African Studies* 7(3): 389 – 412.

Thorsen, D. (2012). *Children Working in the Urban Informal Economy: Evidence from West and Central Africa*. Briefing Paper No. 3. Senegal: UNICEF.

UNICEF.(2002). Poverty and exclusion among urban children. Florence: UN Children's Digest 10, Research Centre

UN Habitat. (2011). State of the Urban Youth 2010/2011: Levelling the playing. Nairobi: UN Habitat.

Zelizer, V.A. (2002). Kids and commerce. *Childhood* **9** (4): 375 – 96.

8

SURVIVAL IN THE MARGINS: *ALMAJIRI* AND WASTE MANAGEMENT IN NORTHERN NIGERIAN URBAN CENTRES

Yusuf Rigasa Abdullahi

Introduction

Almajiri are itinerant male students in the Islamic educational system. The system is driven by low-income rural parents who give out their children (at 4-5 years) in God's name to a trusted Islamic scholar (*Mallam*) for induction into the Islamic way of life. As Muslims are enjoined to travel as far as China in search of knowledge, knowledge-driven migration is well established in Islamic education. Consequently, *Almajiri* abound in all reputable Muslim communities. The institution/system is *al-majirci*, the students are *Almajiri*, the school is *tsangaya* and the teachers are *Mallams*. *Al-majirci* is a comprehensive system of education for learning Islamic principles, values, jurisprudence, theology and ancient science and engineering (see Figure 8.1). No fees are charged by the *Mallams* for their services, as most parents have nothing to give. The *Mallams* are responsible for teaching, counselling and the guidance of *Almajiri,* while the *Almajiri* are responsible for maintaining the *Mallams,* as they have no income. It is believed that on graduation the *Almajiri* will mature into the honest God-fearing men necessary for the growth and development of Muslim societies. Prior to colonial occupation and Western education, *Almajiri* were sponsored by emirs and the local rich. Currently, this patronage is either totally lacking or in steady decline (Abdulmalik 2010). Presently the system survives on the faith and generosity of poor Muslims among whom the *Almajiri* have settled for their scholarly activities and the challenges of life. Most *Mallams* survive on charity from local communities as well as small payments in cash or kind from their students. Income and survival for the *Almajiri* is derived from gifts, begging and petty jobs (see Figure 8.2). Informal waste management activities are readily available and accessible to these scholars, especially in deprived areas of the north.

Several attempts have been made to integrate the *Almajiri* schools into the Western education system, but without success, the most recent being the recent attempt by the federal government to build 400 schools for them (*Leadership* 2012). It is believed that the failure was due to lack of trust, consultation and recognition of the *Almajiri* system as an alternative educational system with a positive contribution to make to growth and the development of society (Lawal 2009).

Figure 8.1: Left: An Almajiri writing on his tablet prior to a class, Kaduna. Right: **Figure 8.2** Almajiri begging at a restaurant in Tegina, Niger state. Note the plastic bottle recovered by one of them.

Waste management in cities of northern Nigeria

The north comprises 19 states and three of the six geopolitical zones of the Nigerian federation. The average poverty rate in the region is 71 per cent (Lawal 2009) and social order is in decline, as traditional rulers are corrupt or support corruption and have lost the respect of their subjects and moral authority to govern them. The solid waste management narratives in urban north share the same features. There is no formal waste sorting or recycling and not all the waste produced is collected. The process is dominated by waste contractors (Abdullahi et al. 2011). In most urban centres, solid waste is disposed of more often in unofficial open dumps and public spaces (Adama 2010). This lowers the quality of the local environment, sometimes with health implications for vulnerable communities. Environmental degradation due to poor waste management can be avoided through environmentally sensitive waste-management techniques, and through the principle of the best practicable options, so that minimisation, reuse, recycling and recovery are employed where feasible to reduce the burden on and need for waste dumps (Wilson et al. 2008)

In northern Nigeria, as in other emerging economies, the major aim of municipal solid waste management is to move up the waste hierarchy through resource recovery and recycling. Informal sector recycling is critical to sustainable waste management (Wilson et al. 2006) by limiting the amount of waste going for final disposal to waste dumps. In most northern cities, only the informal sector engages in resource recovery and recycling (Adama 2010). Waste management in most cities is commercialised. The system is managed by waste contractors who are subsidised by public funds. Selection and monitoring of waste contractors lacks a rigid structure or legal framework, and is based rather on political convenience. Thus, contactors may be paid for services not rendered or services that are slow, irregular or ineffective. The absence of public participation is an impediment to cost and resource recovery, even where such potential exists. Currently, a number of street sweepers and labourers engaged by waste contractors do recover waste resources for sale to the informal recyclers that are emerging in Northern cities.

Effective waste management systems are essential to productive, competitive, healthy and well-functioning cities (Hari 2001). Constitutionally municipal waste disposal and sanitation are the responsibilities of counties or local government councils. However, this tier of government has not performed credibly (Hyuwa 2010). For example, in a survey of waste disposal habits in Kaduna metropolis, Hyuwa,(2000) reported that 38 per cent of the population burn their waste, 37 per cent used the services of the informal sector, 8 per cent engaged private waste contractors and only 3 per cent buried their waste.

As value chains of end-of-waste resources evolve in informal waste management, the engagement of *Almajiri* and the deprived will be critical in job creation and improving the quality of life of low income families. This chapter reviews the role of the informal sector in waste management activities of northern Nigeria's cities and proposes a sustainable solution for the integration of *Almajiri* into mainstream waste management. It is believed that while educational integration has so far failed (Odoh 2012; Oladipo 2012; Paschal 2012; Yajiwa et al. 2012; Zaman 1999), economic assimilation may succeed in the long term.

Almajiri Educational System and the Informal Sector

Almajiri pupils learn the Qur`an under a *Mallam* three times a day without good shelter, food or healthcare and away from their parents, some of whom are hundreds of kilometres away. A pupil's consolation lies in the belief that his suffering in this world is for the reward that awaits him in heaven. He learns how to recite and memorise the holy book in verses and chapters and write them on a wooden tablet called an *allo*. Afterwards, he wipes away the verses he has memorised and is instructed to proceed to the next chapter (*Surah*) as soon as his teacher is satisfied that he has mastered the last one. The duration of his scholarship as an *Almajiri* has no limit but is dependent on his intellectual capability: what matters is for him to memorise the Qur`an. To graduate from the system, he goes through three stages: learning the Qur`an by heart (*Tilawa*), memorising it (*Hafizi*) and perfecting his ability to write the whole Qur`an without error on sheets of paper or a slate (*Darasi*) (Lawal 2009), a timeframe of 20 years.

A survey conducted by a ministerial committee showed that there are more than 9.5 million *Almajiri* in Nigeria. The regional distribution is illustrated in Figure 8.3 below:

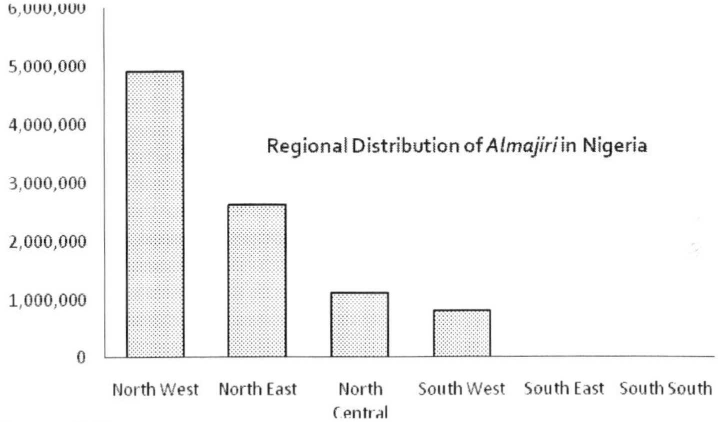

Figure 8.3:
Almajiri in Nigeria (Source: Edet 2011)

According to 2009 statistics from Kano state ministry of education, there were 1.6 million *Almajiri* in 26,000 *tsangaya* (*Almajiri*) schools across the 44 local government areas in the state. Sokoto state has 1.1 million *Almajiri* in 19,167 schools, but the state ministry of religious affairs says this figure is not conclusive, as some villages had been omitted from the tally. Kaduna state has 824,233, while Borno state, by repute a centre of Islamic learning, has 389,048 *Almajiri* pupils. Of this number, Borno indigenes account for 266,160, other states 118,280 and non-Nigerians 4,608 (Lawal 2009). There are 4,464 *Mallams* in the state. These figures may confirm what Aja Nwachukwu, former minister of education, said in April 2008 to the effect that Nigeria accounted for over 11 million of 80 million children worldwide who were outside formal education. Prior to the emergence and dominance of Western education, *Almajiri* were held in high esteem and welcomed, fed and accommodated by host communities and *Mallams* were sponsored by emirs. The pupils worked the farms given to their *Mallams* by traditional rulers, gathered firewood, which they used to read at night. They also assist the families of their *Mallams* by performing domestic chores in return for food, and no pupil was taxed, as the teacher gets his reward from Allah. Currently, most *Almajiri* lack the attributes of past scholars, such as endurance, morals and discipline, and their intellectual capacity is low. As a result, they are highly vulnerable to abuse and criminal activities.

The traditions of Islamic learning emphasise the supremacy of spiritual and moral values over bookish learning, the application of knowledge to guide everyday life and intellectual quest as a lifelong endeavour. Also central to these traditions is a master-disciple relationship characterised by deference to the master, spiritual guidance, moral authority, piety and blessing (Maqsud 1978). These traditions have produced outstanding Islamic scholars, including a number of prolific authors (Mohammed and Khan 1981).Although these ancient traditions are still very much alive, the *tsangaya* schools are facing serious crises of relevance in contemporary Nigeria (Dan-Asabe 1996). As Okoye and Yau (2001) have documented, *tsangaya* have increasingly failed to fulfil their traditional educational mission. Instead of educating their pupils and providing them with the skills and knowledge necessary to function effectively in society as they used to, *tsangaya* schools have so deteriorated that many people regard them as nothing more than breeding grounds for street beggars. Most urban dwellers are either antagonistic to *Almajiri* or engage them in petty jobs and small errands (Lawal 2009). Consequently, the *Almajiri* resent the people around them, who show them no affection, but instead look for opportunities to use them for one purpose or other before assisting them. Many of them have never experienced parental care in their lives.

Informal Sector Solid Waste Management

One important and visibly active group among informal waste management actors in Muslim-dominated areas in the urban north are the *Almajiri*. Male children between the ages of 5 and 10, they are opportunistic waste pickers (Adama 2010). They sort and keep waste, particularly plastics and light metals, from their own homes and from neighbours, and sell it to middlemen in their neighbourhoods. While it is common to see *Almajiri* begging on the streets, the more resourceful among them have taken to waste picking to feed themselves.

As noted earlier, the official waste disposal system is open dumping. Waste from households and factories are dumped at two sites in Kaduna. There is no official sorting, but some of the personnel working for waste-disposal contractors do sort and pick waste. Value pickers are also allowed on to the dump sites to recover end-of-waste resources. These groups of pickers also sell to middlemen (Adama 2010).

In many northern cities, an informal waste sorting system is evolving based on the activities of itinerant buyers of scrap and sorted waste; the barrow boys who collect and dispose of household waste for a fee; and the "*baban bola,*" the so-called masters of refuse dumps who recover valuable items from backyards and local waste dumps. The hard times being experienced by families in deprived areas of northern Nigeria are currently driving the evolution of value chains in the informal waste management sector in the urban north. This sector is dominated by deprived *Almajiri,* orphans, the unemployed and the less privileged. They are characterised by low social status, little or no formal education and a lack of access to the financial instruments needed for a better quality life. The *Almajiri* used to be the largest segment of the community workforce and made significant contributions to the socioeconomic well-being of Muslim communities in northern Nigeria before the advent of white collar jobs. The system produced judges, clerks and experienced administrators in the pre-colonial north and *Almajiri* were the first to be recruited by the colonial government. However, with advent of Western education, they were gradually replaced and abandoned. Consequently, they became victims of negative government policies, public persecution and persistent ethno-religious violence (Abdulmalik 2010). In addition, there has been lees patronage from among the high and mighty in the Muslim societies of the north. These factors have contributed to some of the desperate means of survival adopted by the *Almajiri* to the detriment of public peace and security (Awofeso 2003). Several reports into ethno-religious crises in the north have indicted the *Almajiri* as ready foot soldiers in communal conflicts and terrorism (Lawal 2009).

Solid waste value chains – the evolutionary trend

Solid waste value chains in northern Nigeria are currently evolving. Gradual integration of formal and informal strategies in the management of municipal solid waste will encourage the stabilisation of value chains in waste management (Abdullahi et al. 2013). In the short term, the value of end-of-waste-resources may be low and inadequate for prosperity. However, as the value chain stabilises and lifecycle of each value chain becomes established, the income from the informal sector will increase and the lifestyle of those involved in it will improve. Potential value chains of end-of-waste resources in Nigeria, including the north, are the following:

Plastics

Current value chain actors include value pickers, street sweepers, middlemen (bulk buyers and converters) and end users (e.g., plastic manufacturers). The plastic recycling industry can be considered very informal at the level of value pickers and middlemen. The support markets for the low-density plastic industry are the manufacturers of packaging materials and the producers of sachet water (pure water). Plastic manufacturers have close links with plastic pellet manufacturers through the supply chain between them. Similarly, middlemen supply end-of-waste plastic resources to plastic producers. Value pickers have strong supply chain links with plastic pellet producers, as they provide them with input materials. The value pickers are quite numerous and scattered, so the middlemen are unable to serve all of them satisfactorily. On average, the price offered value pickers by middlemen for recovered plastics is ₦ 20/kg for low density and ₦ 40/kg for high density plastics. Processed pellets are sold at ₦ 100/kg to manufacturers (US$ 1 = ₦ 155).

Metal Scraps

In this case, the main value chain actors are value pickers, scrap agents, dealers (middlemen) and steel rolling mills. Ferrous metal scrap is recycled by the steel mills into iron rods, sheets, pipes and nails for the construction industry. The preference for scrap metal as secondary raw material is because there are limited quantities of iron ore in Nigeria. There are three steel rolling mills in the north, in Katsina, Jos and Kano. Average prices for scrap metal offered by local middlemen are ₦ 20/kg for iron, ₦ 600/kg for copper, ₦ 100/kg for aluminium and ₦ 150/kg for lead. Mechanical engineering workshops that repair motor vehicles and other heavy equipment generate large quantities of ferrous scrap, which feed the scrap market. Scraps from the building industry also support the industry. Currently, the steel rolling mills have supply chain links with the scrap dealers/agents. This linkage is demand driven. Likewise, scrap dealers/agents have supply chain links with value pickers, which operate on a supply and demand basis.

Degradable organic waste

Major value chain actors here are value pickers, compost plant operators and waste contractors. Currently, there is no supporting structure or framework for biogas and composting plants in the north. Consequently, organic and other degradable wastes have little or no value. The large food/fruit markets that traditionally sell unprocessed foodstuffs to households, restaurants and hotels and other commercial eateries are the main suppliers of the compostables. Compost facility operators will have to depend on private waste contractors for a constant supply of compostable material. They will also have to link up with distributors of agricultural inputs and supplies.

Role of *Almajiri* in the evolution of an Integrated Solid Waste Management Strategy (ISWMS) in Northern Nigeria

Almajiri as waste managers

The *Almajiri* as value pickers can integrate door-to-door waste collection and source-separation with the transportation, processing and marketing of recovered resources in a way that is cost effective and eco-friendly. *Almajiri* could lower the cost of waste transportation to local waste collection points (LWCPs), communal bins and waste dumps by operating a "bring system" for end-of-waste resources in conjunction with local residents, in which the waste dump can serve as a recycling market as well as a civic amenity. Recycling markets are central to the proposed strategy, as shown in Figure 8.4. They will act as all inclusive meeting points for the interaction of the *Almajiri* on one hand and the state's ministry of environment, solid waste management agencies (SWMAs), local council officials, waste contractors, industries, residents and research institutions on the other.

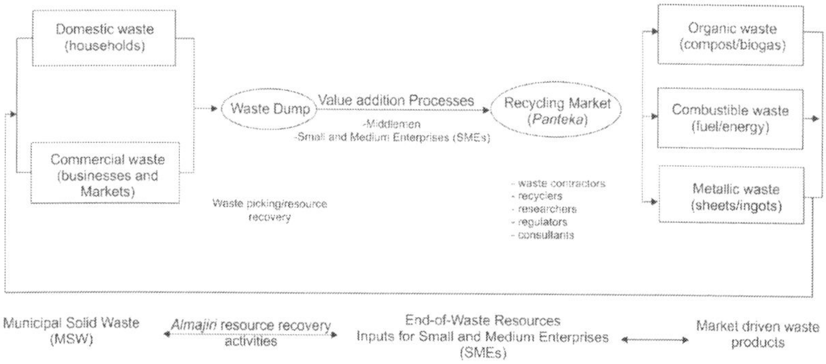

Figure 8.4:
Framework for integrating *Almajiri* as value pickers into the evolving waste management services of northern Nigeria (*Source*: the author)

The ISWMS could become sustainable where the value of recovered resources offsets the transport costs. Waste dumps and recycling markets could be used for small-scale anaerobic digestion and composting processes as a means to recover value from biodegradable waste. Anaerobic digestion and composting are sustainable techniques for the production of energy (in the form of biogas) and soil fertiliser (Abdullahi et al. 2008). Efficient recycling and composting programmes could save 78.1 per cent of waste management costs and 79.5 per cent in landfill avoidance costs (Agunwamba 2003). Sorted combustibles waste can be processed into waste fuel for incineration or for use in domestic cooking. Value can also be recovered from sorted metals and non-combustible end-of-waste resources, which can serve as raw materials for small and medium industries. In the proposed ISWMS, the recycling market is critical in optimising value/cost recovery from the sale of recyclables and other end-of-waste resources, as the activities of middlemen will be greatly reduced. The *Almajiri* could also aid with waste awareness and public participation as a consequence of their activities.

Waste contractors

The use of contractors can be sustainable where their activities are effectively regulated and restricted to areas where waste management costs can be fully recovered. Waste collection by contractor should therefore be restricted to affluent residential districts, public institutions and corporations that can afford the user fees. Contractors can also be engaged in transporting waste from LWCPs and communal bins to waste dumps. As the end-of-waste resources grow in value and volume, waste transportation should be left to the informal sector, to be monitored by states assisted by local government councils. General sanitation in cities requires some formal organisation, which only state governments, SWMAs and local councils can afford. However, the *Almajiri* can be sub-contracted by waste contractors or SWMAs in the sanitation of low-income districts

Local councils

Under the proposed strategy, local councils will be responsible for providing and maintaining LWCPs, waste dumps and recycling markets, where a fee could be charged for space and onsite amenities, but not for waste disposal. The councils can aid informal sector activities at waste dumps and recycling markets through onsite training, the sale or lease of the simple tools required to separate and transport recovered recyclables and overall maintenance and enforcement of sanitation and public health and safety measures.

Households, commercial and institutional waste producers

Households should be made responsible for delivering waste to LWCPs, which could be open sites or communal bins. Where this is not possible, the *Almajiri* can collect and deliver the waste free of charge or for a token fee, depending on the potential of the end-of-waste resources. Commercial and institutional waste producers can afford prescribed fees and therefore be effectively serviced by contractors. However, contractors should be appointed based on their competence in resource recovery. Where the use of contractors is not cost-effective, the proliferation of private waste contractors and *Almajiri* should be encouraged. Both private waste contractors and *Almajiri* should engage waste producers in resource recovery schemes, and where households are actively engaged in recovering recyclables from their own waste, their efforts should be rewarded.

Solid Waste Management Agencies

SWMAs are supposedly the bodies responsible for regulating waste management in urban centres of the north. In all northern states, solid/municipal waste is managed by ministries of the environment with little input from the public, consultants, waste contractors and other stakeholders. Political interference and the lack of a functional regulatory framework with clearly defined roles for each agency are undermining the efficiency of existing strategies. SWMAs should be given the power to appoint and monitor waste consultants and contractors, as well as to regulate informal waste management activities. They could formally engage the *Almajiri* under the guidance of the ministry of environment based on the objectives of the chief

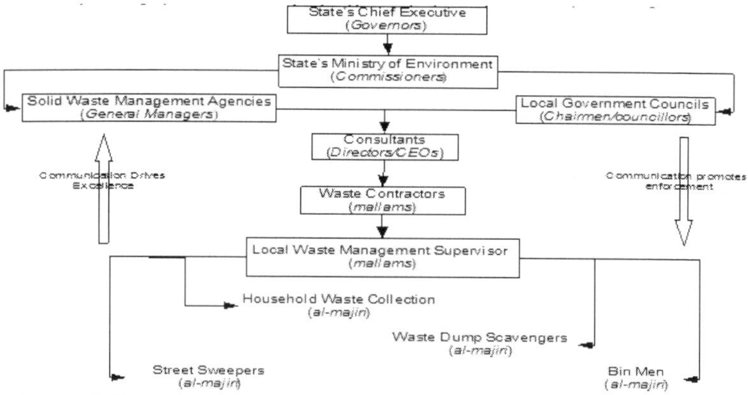

Figure 8.5:
Regulatory Framework for the proposed ISWMS
(*Source*: the author)

proposed in Figure 8.5 will drive policy formulation and promote enforcement through effective communication with all stakeholders.

The agencies may also require the services of technical consultants to train stakeholders and clearly define their roles and duties. Contractor performance should be evaluated based on cost and the value of recovered end-of-waste resources. The agencies should be responsible for setting new targets, procuring new techniques and equipment as well as capacity building and regulation of health and safety issues. SWMAs should encourage the evolution of sustainable partnerships between the *Almajiri*, research institutions and consultants and local councils to optimise the value, volume and diversity of end-of-waste resources. This may require substantial financial and technical inputs. SWMAs should also be responsible for raising awareness of environmental issues, including solid waste management. This can be achieved through the promotion of environmental education, environmental clubs in schools as well as through print and electronic media. Attempts should be made to integrate local traditional values to encourage as well shape public participation in waste management.

Conclusions

Urban centres of northern Nigeria are currently undergoing rapid expansion, which is placing increased pressure on scarce public resources and infrastructure, especially those relating to solid waste management. Current action plans are working but are not sustainable. The *Almajiri* have long been regarded as backward and ineffective components of Municipal Solid Waste (MSW) management in the cities. This chapter has demonstrated that *Almajiri* can be used as the fulcrum for resource-recovery based ISWMS in cities where cost-recovery based strategies have failed, are ineffective or are unsustainable. The proposed strategy is all-inclusive and requires little or no capital investment, but provides greater economic, social and environmental benefits to all stakeholders. The strategy emphasises the use of *Almajiri* as value pickers and has as a key element recycling markets that can serve as centres for value recovery, training, awareness-raising and some formal regulation. The performance and sustainability of ISWMS depend on the presence and functionality of the informal sector. The integration of solid waste management with land-use and public health initiatives will be required to optimise the strategy's benefits.

Recommendations

This chapter has demonstrated that solid waste value chains are just evolving in northern Nigeria. The *Almajiri*, as a deprived segment of the society, are already engaged in informal waste management activities. The *Almajiri* in collaboration with other informal waste management actors can add value to waste management services in population centres of northern Nigeria and could contribute significantly to the evolution of sustainable waste management in Nigeria. The recognition of *Almajiri* as the equal partners of waste contractors, local councils and SWMAs in the management of municipal solid waste is strongly recommended.

References

Abdullahi, Y.A., K.G. Jiriko, and J.C. Akunna (2011).*The role of Informal sector in the delivery of sustainable waste management services in Kaduna Metropolis*. In Proceedings of the 26th International Conference on Solid Waste Technology and Management, Philadelphia, 27-30 March.

Abdulmalik, I. (2010). Re-Almajiri-Education, *The NATION Newspaper*, 5 April. Available online at http://thenationonlineng.net/web2/articles/41947/1/Re-Almajiri-Education/Page1.html.Accessed 11 August 2011.

Adama, O. (2010). *Beyond Poverty: An insight into the informal solid waste recycling sector in Kaduna, Nigeria*. Available online from: http://nai.diva-portal.org/smash/record.jsf?pid=diva2:413279. Accessed 11 August 2011.

Agunwamba, J.C. (2003). Analysis of scavenger's activities and recycling in some cities in Nigeria. *Environmental Management* **32**(1): 116 – 27.

Arif, B. and H. Abiyoga (1991).Private sector participation in the urban solid waste management in Indonesia. Paper presented at the International Seminar on Policy Response for Improving Solid Waste Management in an Asia Metropolis, Bandung, Indonesia.

Awofeso, N., J. Ritchie and P. Degeling (2003). The *Al-majiris* heritage and the threat of non-State terrorism in northern Nigeria: Lessons from Central Asia and Pakistan. *Studies in Conflict and Terrorism***26**: 309 – 23.

Daily Trust (2011). Minister: 68m Nigerian youths out of job, *Daily Trust Newspaper*, 9 August, p. 3.

Dan-Asabe, A.U. (1996.) Islam and History of Learning in Katsina State from the Jihad (1807) to the Colonial Conquest (1903): The Case of the School of Tsohuwar Kasuwa, Katsina City, Tambari. *Kano Journal of Education* **3**(1): 72 – 8.

Edet, B. (2011). Towards a harmonized *Al-majira*i education curriculum, *Daily Trust Newspaper*, 30 June, p. 47.

Hyuwa, B.A. (2010). *Solid Waste Management in Nigeria: An outline*. Paper presented at the Annual National Conference on Environmental Challenges: Local Action Plans, Nigerian Environmental Society, Kaduna State University, 3 – 5 June.

International Solid Waste Association (ISWA) (2002).*Industry as a Partner for Sustainable development: Waste Management*. Copenhagen: International Solid Waste Association.

Lawal, A. (2009). *Almajiri* – Breeding of Future Terrorists, *Tell Magazine*, 27 December. Available online from: http://www.citizensfornigeria.com/index.php?option=com_content&task=view&id=262. Accessed 11 August 2011.

Leadership (2012). F.G. Approves Establishment of 400 *Al-Majiri* Schools, *The Leadership Newspaper*, 17 February. Available online from: http://www.leadership.ng/nga/articles/16667/2012/02/17/fg_approves_establishment_400_almajiri_schools.html. Accessed20 August 2012.

Maqsud, M. (1978).Moral Education in Islamic Schools in Northern Nigeria. *Kano Studies, New Series* **1**(3): 53 – 60.

Mohammed, A. and M.B. Khan (1981). From Cradle to Grave: The Contribution of the Ulama to Education in Nigeria. *Kano Studies New Series* **2**(2):110 – 28.

Medina, M. and M. Dows (2000). A short history of scavenging, *Comparative Civilizations Review,* 23-45

Odoh, P. (2012). *The need to remodel al-majiri system in Nigeria.* Available online from: http://desertherald.com/?p=1434. Accessed 20 August 2012.

Okoye, B.A. and A.A. Yau (2001).The *Almajiranci* System and Social Realities.*Islamic Culture* **75**(3): 85 – 103.

Oladipo, A. (2012). *The al-majiri syndrome: searching for lasting solution.* Available online from: http://tribune.com.ng/index.php/politics/39561-the-almajiri-syndrome-searching-for-lasting-solutions. Accessed 20 August 2012.

Paschal, O. (2012).*The fallacies of al-majiri project.* Available online from: http://www.punchng.com/opinion/the-fallacies-of-al-majiri-project/. Accessed 20 August 2012.

Wilson, D.C., A.O. Araba, K. Chinwah and C.R. Cheeseman (2009). Building recycling rates through informal sector. *Waste Management,* **29** (2) 629-635.

Wilson, D.C., C. Velis, and C. Cheeseman (2006).Role of informal sector recycling in waste management in developing countries. *Habitat International***30**: 797 – 808

Yajiwa, H.K., S.M. Gusau, A. De-Don, M.I.J. Afolagbe and F. Bashir (2012). *Why the al-majiri syndrome will persist.* Available online from: http://elombah.com/index.php/latest-news/10537-why-the-almajiri-syndrome-persists-group. Accessed on 20 August 2012.

Zaman, M.Q. (1999). Religious Education and the Rhetoric of Reform: The Madrasa in British India and Pakistan. *Comparative Studies in Society and History* **41**(2): 294 – 323.

PART FOUR
RELATIONS WITH THE STATE

9

TOWARDS GOVERNMENT RECOGNITION OF AND SUPPORT FOR WASTE PICKERS IN NIGERIA

Anthony C. Okoye

Introduction

Municipal solid waste management is one of the most crucial environmental challenges facing developing countries. In most countries, the urban population is growing faster than urban authorities' capacity to collect and dispose of waste, leading to inadequate waste management. The in-migration of people with different cultural backgrounds and from different localities in search of greener pastures has virtually turned cities into a "no man's land," so that fundamentally environmental maintenance becomes no man's business (Omuta 1988). Most developing countries, including Nigeria, have solid waste management problems different from those in developed and industrialised countries in the terms of composition, density, political and economic framework, access to waste collection, awareness and official and public attitudes (Ogwueleka 2009)

In Nigeria, waste management is inadequate and ineffective in many urban and rural areas, as the federal environmental protection agency acknowledged in its report to the United Nations Conference on Environment and Development (FEPA 1992). The main version of the national assessment report for the World Summit in Sustainable Development (WSSD) declared that Nigerian cities are considered among the dirtiest in the world (WSSD 2002). The mountainous heaps of solid waste that deface these cities are a clear indication that the solid waste management system is unsustainable and grossly inefficient.

Recycling is a key component of modern waste reduction and is the third element in the Reduce, Reuse and Recycle waste hierarchy (Wikipedia n.d.). Until recently, there were no formal recycling or resource recovery programmes in Nigeria. Recycling activities, which are carried out by the informal sector, are often driven by poverty and the desire to earn a living. Waste scavengers/pickers are part of an informal waste recycling sector that also consists of cart pushers, cart builders, resource merchants and recyclers (Adebola 2006a). Recently, urban waste management in Nigeria, especially the role of the informal sector, has attracted increasing attention from researchers and policymakers (Nzeadibe and Ajaero 2011; Okoye 2013). In all these studies, the development of informal sector capacity for improved solid waste management and the need to support the activities of the sector have been largely ignored (Nzeadibe and Ajaero 2011). This is despite the fact that waste pickers are the major recyclers in Nigeria. They collect, sort, clean and transport the recyclable fraction of wastes to recyclers. The non-recognition and lack of support for these waste pickers by government is what this chapter seeks to discuss.

Potential of Recycling in Nigeria

Recycling involves processing used materials into new products to prevent the waste of potentially useful materials and reduce consumption of unprocessed raw materials, as well as reducing energy use, air pollution from incineration and water pollution from landfills by moderating the need for conventional waste disposal. It also entails lower greenhouse gas emissions as compared to production using unprocessed raw materials (PM's advisor hails recycling as climate change action 2013; League of Women Voters 1993). Recycling provides a livelihood to approximately 15 million worldwide, or 1 per cent of the urban population in the developing world (Medina 2008). Thus, it is preferable to waste disposal on virtually every count: greenhouse gas emissions, release of toxins, employment, social inclusion, sustainability and public expenditure (Tangri 2010). It is certainly a tremendously cost effective method of achieving emission reductions.

Waste recycling is untapped business in Nigeria. Given the fact that the country generates about 25 million tons of waste annually (Ogwueleka 2009), waste recycling and composting should be encouraged, since this approach is considered to be the right way to achieve sustainability. Efficient recycling and composting could save 18.6 per cent in waste management costs and 57.7 per cent in landfill costs, (WSSD 2002; Ogwueleka 2009), not to mention energy costs compared to processing raw materials, and could prolong the life of dump sites and landfills. In Onitsha for instance, 40 per cent of artisans and small-scale industries obtain 48 per cent of their raw materials from scavengers (Ogwueleka 2009). It was reported in *Tell* magazine (Eni 2009) that 30 members of the Tinsmith and Stoves Traders Association manufacture industrial stoves, ovens and coal pots from iron and metal scraps sourced from dump sites. These stoves sell for between ₦ 3,500 and ₦ 10,000 depending on size. An investigation in Anambra state revealed that on average 19 full trailer loads of scavenged metal left the state per day to metal recycling plants elsewhere in Nigeria (Muoghalu and Okoye, 2010). One can then imagine the volume of such materials leaving all the other states. All these examples show the potential of recycling in the country, despite limited government support.

Plastic waste constitutes a very high percentage of both domestic and industrial waste in Nigeria. This can be recycled into various end products. Used pure water sachets constitute an unquantifiable nuisance in Nigeria. This problem will continue because of the importance of the sachets in making drinking water available to the common man. Recycling such waste could create many job opportunities for Nigerians. On the other hand, e-waste disposal has remained a major problem in developing countries, where electronic products are likely to be used and repaired before final disposal. The advent of used (*Tokunbo*) electronics and allied products in Nigeria has led to an increase in the e-waste generated in the country. Currently, waste pickers collect these items and isolate the recyclable components. However, they do this in a way inimical to their health. If this activity could be harnessed in appropriately, it could be a tremendous source of jobs and revenue.

The great majority of wastes generated in Nigeria are organic. In the strict sense, organic wastes may not be regarded as recyclable, but they can be used as feedstock for another item, for instance, organic fertiliser. In Lagos, LAWMA (Lagos State Waste Management Authority) had set up three recycling locations, including Odogunyan in Ikorodu, where it sited the Waste to Compost plant, which accounts for 8 per cent of LAWMA's recycled products. Compost Plus is the fertiliser produced through the recycling efforts there. Earthcare Incorporated, the private sector investor partnering with government, churns out 250 bags of this organic compost daily. The product comes out like the fertile black soil common in deep uncultivated forest lands and it has no odour or chemical additives. Compost Plus is derived from vegetables, food waste and other wastes from markets. It is used as manure for the trees and flowers associated with the beautification efforts of the Lagos state government. This experience should be replicated in, and beyond, all seven cities where the federal government instituted the national integrated municipal solid waste management intervention programme, but this has yet to happen. This is an aspect of resource recovery that Nigerians have yet to exploit fully.

Paper is one of the major wastes generated in Nigeria, and can be loosely classified into two classes, "clean" and "soiled." Waste paper that has been in contact with garbage or otherwise dirtied falls into the latter category (Golueke and Diaz 1991). Demand for soiled waste paper is currently negligible, except as fuel for waste-to-energy plants. It is tolerated in compost plants and will be used in biogas production and fibre recovery plants (Golueke and Diaz 1991).

Nigerians generate tremendous amounts of solid waste that could be reused or recycled. There is no gainsaying the fact that the gap between rich and poor could be narrowed in Nigeria by mainstreaming and implementing recycling programmes in environmental policies. Waste pickers all over the country should be supported and empowered by the tiers of government in view of waste picking's numerous benefits, including poverty reduction, job creation and environmental sustainability.

Government and Waste picking globally

The relationship between waste picking or scavenging systems and the authorities varies markedly from country to country. Government authorities in most countries do not appreciate the roles scavengers play in municipal waste management (Nas and Jaffe 2004). Several studies have reiterated the need for the formal sector to partner with informal waste workers for more effective management of waste (Adebola 2006b; Nzeadibe and Ajaero 2011; Okoye 2013; Arimah 2001; Dias 2011; Akintoye 2008; Olufayo and Omotosho 2007; Kuria and Muasya 2010).

One major element in strengthening informal sector integration involves raising awareness among political decision-makers about the work and possible role of the informal sector in an appropriate waste management system, as well as about the important aspects to consider in the integration process (Kuria and Muasya 2010). Government officials sometimes develop relationships of exploitation, mutual profit and mutual assistance – that is, relationships of political clientelism – with scavengers (Medina 2007). Scavenging should be viewed as a basis for further development, instead of as an impediment to development, as governments still often implicitly or explicitly assume (Nas and Jaffe 2004).

The relationships of various national governments with scavengers were highlighted by Medina (2007). He noted that in Mexico, recovery of recyclable materials is carried out informally by scavengers. However, Mexican authorities generally do not see that scavenging has a place in modern solid waste management. Nor do most Mexican mayors and local government officials perceive recycling as something cities should support, even though the informal waste management sector in Mexico is quite large and has a significant social, economic and environmental impact.

The first world scavenger cooperative movement was formed in Colombia in the 1980s. During the 1980s and 1990s, the Colombian national government played a limited role in supporting the development of scavenger co-ops. In 1999 the Colombian senate passed Law 511 legalising scavenging, and the president issued Decree 2395 a year later, which greatly improved scavenging and recycling activities in Colombia (Medina, 2007)

In Argentina, scavengers are commonly called *cartonero*s, and have a significant social, economic and environmental impact in the country. The authorities' earlier repressive actions towards *cartoneros* provoked sympathetic support from the public. Thereafter, the Buenos Aires legislature approved Law 992, which legalised and supported *cartonero* activities in the city, and led to the creation of the Programa de Recuperadores Urbanos. Its primary role is to manage the register of *cartonero* activities and to work towards an integrated waste management approach that incorporates the scavengers (Medina, 2007).

The literature on scavenging reveals that scavengers in many countries are better off than their Indian counterparts, who are discriminated against by society, ascribed the lowest status, work under hazardous conditions and face repressive authorities. Indian authorities at all levels have largely ignored scavenging, despite the recommendations of a Supreme Court committee in 1999 that recognised the work of the informal recycling sector and recommended assistance to scavengers to organise themselves to participate in source-separation programmes. Not deterred by official indifference and negligence, the scavengers went ahead and created SWACHH, an alliance of waste pickers in India (Medina, 2007). It is noteworthy that the economic impact of scavenging activities in India exceeds US$ 280 million per year (Medina, 2007).

The inference that can be drawn from the foregoing is that support to scavengers or waste pickers in terms of organising them, financing, training and health facilities significantly improves the efficacy of their waste-picking activities, their psychological well being and the aesthetic character of the country, and alleviates poverty by creating many jobs. There is a need to review relevant municipal bylaws in order to accommodate the waste picking sector within the larger framework of waste management (Samson 2010)

Review of Waste picking in Nigeria

In Nigeria, as in many other countries, scavenging is a reality. Waste pickers roam the streets and waste dumps to recover the recyclable items that many industries rely on for their business activities. In Kano, it is common to see waste pickers roaming the dumps with sacks and sticks. Mainly youths, these scavengers are not organised in any formal way (Muktar 2009). They expose themselves to all sorts of health hazards and do not get commensurate pay. According to Muktar (2009), they make ₦ 12,000 a month on average. In Abuja, the waste pickers who go from house to house collecting recyclables (itinerant buyers) are also not organised, but those that operate at the dumps have an association to regulate their activities (Ezeah et al. 2009). Studies in western Nigeria reveal that waste pickers are involved in both onsite and off-site materials recovery (Adebola 2006a; Olufayo and Omotosho 2007). In some cases, they have associations, but in others, not. It is noteworthy that the sole aim of the associations is not financial aid but assistance with the other problems often encountered while collecting recyclables.

Studies by Muoghalu and Okoye (2010) reveal that waste pickers in the eastern part of Nigeria have a union, but that it does not regulate their interactions with middlemen or environmental institutions: the union is for the protection of members who get into trouble. They have no relationship with the state environmental protection agency or waste management authority. On the hazards of their business in terms of illness, pickers responded that their regular illnesses were malaria, physical injuries sustained while ransacking waste heaps and prolonged coughs. Their greatest complaint is physical injuries, which they regard as normal for their type of business. This is to be expected as the only equipment they have is their hands and sticks, with which they prod the heaps of waste. None of them have personal protective equipment such as boots, gloves, masks, hand forks/rake, caps or aprons. In depth interviews put their daily income at between ₦ 1,500 and ₦ 2,000 daily. This means that their monthly income ranges from ₦39, 000 to ₦ 52, 000 (Muoghalu and Okoye 2010).

The major challenges and constraints facing waste picking in Nigeria can be summarised as follows:

- Occupational and health hazards due mainly to non-use of personal protective equipment and lack of other facilities that would make the job easier;
- Negative perceptions by government, the public and waste pickers themselves, which undermines their efforts and limits the scope of coverage;
- Lack of capital to buy the recyclables and lack of means to convey them; and
- Exploitation by middlemen

On the whole, waste pickers are responsible for more than 90 per cent of the recycling of solid wastes in Nigeria, thereby contributing immensely to reducing greenhouse gas emissions, cleanliness of the country and prices of raw materials. Their recycling activities result in less demand for virgin raw materials, whose extraction, transport and processing are a major source of greenhouse gas emissions. However, their output and benefit to the country would increase significantly if government recognised their role and gave them administrative, legal and financial backing.

Way Forward

Nigeria has a serious solid waste management problem. Urban population is outstripping the authorities' capacity for waste collection and disposal, leading to inadequate waste management. Recycling is a key component in the waste hierarchy and provides employment to millions of people, especially in developing nations, thereby alleviating abject poverty. Government recognition of and support for waste pickers is crucial. Nigerian waste generation is on the increase at an estimated rate of about 0.5 – 0.7 per cent per annum, with current figures ranging from 0.4 to 0.8 ton per capita per annum. Waste complexity is also increasing, with biodegradables currently accounting for over 50 per cent of waste. This amounts to over 50 million tons per annum in *average waste burden* on a state with less than 10 per cent *waste management capacity*. Simply put, this implies there is a goldmine to be discovered in one of the environmental challenges facing Nigeria.

The objectives and strategies mapped out in the policy guidelines on solid waste management Nigeria include:

Objectives

1. To minimise waste generation and promote sorting at source, reuse, recycling and energy recovery; and
2. To generate employment opportunities, improve the standard of living and thus reduce poverty

Strategies

1. To evolve and promote appropriate technologies for recycling waste components such as bottles, glass, metals, paper, plastic and organic matter; and
2. To foster the establishment of small-scale waste recycling plants at household and community levels to source and convert recyclable waste

Nowhere in the guidelines is the informal sector or waste pickers mentioned, which implies that government not only ignores them but also has no plan to support them for now. Regarding the first objective above, government has made no frantic efforts to implement it. The first strategy will make room for the second objective, but government implementation efforts have been inadequate and have not yielded fruit. For instance, only the federal government and a few states have started to venture into resource recovery and recycling. A resource recovery and bailing centre was constructed in Abuja at the cost of ₦ 275.9 million. The plant, according to a former minister of the environment, would be used to sort waste materials and bale the recyclables for onward transfer. The last I heard about the place was that it had been vandalised and till today is not operational. If this huge amount were used to empower some 1,100 waste pickers, each of them would get ₦ 250,000 and by the end of a year there is every chance each would have made double the money and could return the seed funds to set up a resource recovery centre in the federal capital territory. If this approach were extended to waste pickers in the six geopolitical zones, the government would be able to set up such a centre in each zone. The implication of this is that by using both strategies, government would have been better able to achieve the second objective, namely generating employment opportunities, improving living standards and reducing poverty.

Recently, the federal government has taken a bold step towards solving this problem through privatisation. It has instituted a national integrated municipal solid waste management programme in seven cities, namely Maiduguri, Kano, Kaduna, Onitsha, Uyo, Ota and Lagos. Prior to this, however, it had instituted the programme in Aba and Ibadan. However, none of the contractors have established the facilities for reasons ranging from politics to incompetence to lack of infrastructure. Thus, it seems that recycling activities in Nigeria so far rest with waste pickers across the country. In Nigeria, hundreds of trailer-loads of scrap metal transported for recycling were the result of the efforts of these unrecognised waste managers. The waste pickers are often at the mercy of middlemen, who pay them a fraction of market prices. The "slavery status" of this segment could be transformed into the status of "upcoming entrepreneurs" if government would give them support. Below are the roles government can play to boost recycling activities and make the efforts of waste pickers more effective, while also creating job opportunities in line with the Millennium Development Goals:

- Government should integrate waste pickers into waste management policy, because they function as the actual recycling system and have the required skills to sort recyclables
- Government can create national educational and training programmes for waste pickers and incorporate them as waste collectors or street beautifiers. This will make their job of isolating recyclables from the wastes easier
- To free waste pickers from exploitation by middlemen, government can legalise their activities and make it compulsory for all of them to be registered, and provide them with personal protective equipment and tricycles as well as a good welfare package in recognition of their contribution to the economy and society
- Government can ensure that waste pickers' output is bought from them at a good rate

- Government can facilitate the organisation and formal recognition of informal waste workers through identity cards, co-operatives, associations, etc.
- Government can help waste pickers gain access to credit facilities, which would lead to a more balanced income among organised waste pickers
- Government can grant waste pickers sufficient space within urban areas to get their businesses more organised

These and other welfare packages will not only make waste picking jobs more attractive to even graduates, but also create tremendous employment and wealth and, in the long term, a clean environment.

Conclusion

This chapter has shown that waste pickers are the drivers of the recycling system in Nigeria. The efforts made so far by government indicate that they appreciate the need for recycling. The efforts are yet to yield meaningful results, because their strategies do not factor in the waste pickers and their roles. The only way forward is to mainstream waste pickers into waste management policy and encourage them.

References

Adebola, O.O. (2006a). *The Role of Informal Private Sector in Integrated Solid Waste Management (ISWM) in Lagos, Nigeria - A Developing Country.* Proceedings of the 21st International Conference on Solid Waste Technology and Management, Philadelphia.

Adebola, O.O. (2006b). The Role of Informal Private Sector in Integrated Solid Waste Management in the Achievement of the Millennium Development Goals in Lagos Nigeria. Paper presented at the Solid Waste, Health and Millennium Development Goals, CWG-WASH Workshop in Kolkata, India.

Akintoye, I.R. (2008). Reducing Unemployment through the Informal Sector: A Case Study of Nigeria. *European Journal of Economics, Finance and Administrative Services.* Available at http://www.eurojournalsn.com.Accessed on 12th December, 2012.

Arimah, B.C. (2001). Nature and determinants of the linkages between informal and formal sector enterprises in Nigeria.*African Development Review* **13** (1)

Dias, S. (2011). Recycling Poverty and Environment: Creativity and Integration of the Informal Sector. Available at *www.institut.veolia.org/.../DIAS_ST4.1_Creativite_et_integration_du_se...* Accessed 14 December 2012.

Ezeah C., C.I. Roberts, G. Watkin, P.S. Philips and A. Odunfa (2009). *An Investigation of Seasonal Variation in Municipal Solid Waste Compostion in Tropical Urban Environment: A Case Study of Abuja.*Proceedings of the 24th Conference on Solid Waste Technology and Management, Philadelphia.

Federal Ministry of Environment (2005). Policy Guidelines on Solid Waste Management, Nigeria.

FEPA (1992).*Transition to Sustainable Development in Nigeria. Report for the United Nations Conference on Environment and Development.* Lagos: Academy Press.

Golueke, C.G. and L.F. Diaz (1991).Potential Useful Products from Solid Waste. *Waste Management and Research* **9**: 415 – 23.

Eni, H.(2009).Nigerian Garbage Millionaires. *Tell Magazine,* 25 May.

Kuria, D. and Muasya R. (2010). Waste Pickers and Organizations Supporting Waste Pickers in Kenya. Technical Report Submitted to Women in Informal Employment Globalizing and Organizing (WIEGO).

League of Women Voters (1993).*The Garbage Primer*. New York: Lyons and Burford, pp. 35–72.

Medina M. (2007).*The World's Scavengers: Salvaging for Sustainable Consumption and Production*. Lanham MD: AltaMira Press.

Medina, M. (2008). The informal recycling sector in developing countries: Organizing waste pickers to enhance their impact. Gridlines No. 44

Muktar, M. (2009). The Youth and Waste Scavenging in Nigeria: Implications for Socio-economic and Health Hazards. Available at http//www.gemji.com/article8000/news58971.html.Accessed 10 February 2011.

Muoghalu, L.N. and A.C. Okoye (2010). Solid Waste Management and economic development: A preliminary report on scavenging in Anambra state. *Tropical Built Environment Journal* **1**(1): 73 – 82.

Nas, P.J.M. and R. Jaffe (2004). Informal Waste Management: Shifting the Focus from Problem to Potential. *Environment, Development and Sustainability* **6**(3): 337 – 53.

Nzeadibe, T.C. and C.K. Ajaero (2011). Development impact of advocacy initiatives in solid waste management in Nigeria. *Environment, Development and Sustainability* **13**(1): 163 –77. DOI 10.1007/s10668-010-9254-8.

Ogwueleka, T.C. (2009). Municipal solid Waste Characteristics and Management in Nigeria. *Iran Journal of Environmental Health, Science and Engineering* **6**(3): 173 – 80.

Okoye A.C. (2013). The Roles of Government in Waste Management. Paper presented at the UNDP-Assisted Stakeholders Workshop on the Development of Wastes to Wealth Strategic Plans in Delta State, April,2011.

Olufayo, O. and B.J. Omotosho (2007).Waste Scavenging as a Means of Livelihood in Southwestern Nigeria. *Pakistan Journal of Social Sciences* **4**: 141 – 6.

Omuta, G.E.D. (1988). Urban Solid Waste Generation and Management: Towards an Environmental Sanitation Policy. In Environmental Issues and Management in Nigeria Development, P.O. Sade and O. Odemerho (eds). Ibadan: Evans Brothers.

PM's advisor hails recycling as climate change action. Available at ".http://www.letsrecycle.com/do/ecco.py/view_item?lis-tid=37&listcatid=231&listitemid=8155&section=legislation. Accessed on 3 February 2013.

Samson, M. (2010). Reclaiming Reusable and Recyclable Materials in Africa: A Critical Review of English Language Literature. WIEGO Working Paper (Urban Policies) No. 16. Cambridge, MA: WIEGO. Available at http://wiego.org/sites/wiego.org/files/publications/files/Samson_WIEGO_WP16.pdf. Accessed on 3 February 2013.

Tangri, N. (2010). Respect for Recyclers: Protecting the Climate through Zero Waste. Global Alliance for Incineration alternatives. Available at www.noburn.org.

Wikipedia (n.d.): Recycling. Available from http://en.wikipedia.org/wiki/Recycliong#cite_note-0.Accessed 12 July 2013.

World Summit on Sustainable Development (WSSD) (2002).*Sustainable Development in Nigeria: Ten years after RIO (UNCED), National Main Report*. Lagos,Nation House Press.

10

INFORMAL SECTOR WASTE RECYCLING IN ABUJA: PROBLEMS AND PROSPECTS

Emmanuel Chukwuocha

Introduction

The city of Abuja was established on 5 February 1976 and provided an opportunity to avoid the unplanned development of grossly inadequate infrastructural facilities for water, waste management and sanitation, electricity and transportation that characterised urban areas such as Lagos. Thus, the Abuja master plan took the issue of waste management very seriously and made elaborate provisions for its management and development. Conscious of mistakes of the past in most of the older Nigerian cities and towns, which evolved without development plans, a comprehensive waste management scheme was proposed for Abuja (Federal Republic of Nigeria 1979). However, during the physical development stages, the Abuja master plan was not followed and account was not taken of informal sector operators, such as waste material collectors, and their role in the waste management process. The informal sector has today become an important part of Abuja's waste management system. This chapter examines the roles of informal sector operators, the problems they face and measures to improve their performance in waste management. These measures are expected to form the basis for setting up and operating a sustainable waste management and recycling system in the city.

Methodology

This chapter is based mainly on the author's practical experience in managing wastes in Abuja for more than 25 years, both as a regulator and as a private sector operator. It also includes interactions with colleagues and associates in waste management who face similar challenges. Key informant interviews, site visits and desktop reviews were used to obtain the information.

The city of Abuja is zoned into five areas. Each has not less than five major informal waste dumps /recovery centres. The population at these dumps includes the site chairmen, representatives of the groups of waste collectors working there and the agents of the dealers in major recovered materials. A total of 125 persons interviewed, five persons at each dump site or 25 persons from each area, mostly in Hausa through an interpreter, since most of these people do not speak good English. The exercise took a total of five days: two for planning and orientation, two for fieldwork and one for assessment and analysis. A team comprising two personnel and a leader was used for the fieldwork.

Abuja Informal Recycling Activities and Structures

The Abuja master plan provided for a network of three depot facilities and two sanitary landfills for the entire city and for about two hectares of land to be reserved for a group of 60,000 inhabitants to aid solid waste processes and their management (Federal Republic of Nigeria 1979: 210 – 211) .However, there is no formal waste recycling in place in Abuja (Ezeah 2012; Nzeadibe et al. 2012). Wastes are disposed in one stream without any segregation at source. The envisaged infrastructural facilities that should have aided waste recycling, such as waste material recovery facilities, recycling plants, compost plants, etc. are not in place in line with the master plan. Abuja is an administrative city, so most of the wastes are taken to more industrialised towns like Kano, Kaduna, Jebba, Lagos and Jos.

Waste material collectors play a major role in solid waste management, especially in reducing the volume of waste that is disposed of. However, the rudimentary methods used in their activities reduce the efficiency of waste management. These waste collectors and used material merchants are sometimes perceived to be a hindrance to waste management operations, as they scatter waste around storage containers and litter the environment (Nana 2010). Scavenging activities prior to routine collection includes breaking open garbage bags to gain access to their contents or indiscriminately removing materials from waste receptacles and leaving them littered all over the place (Chukwuocha 2004).Collectors operate at dumpsites/waste dumps and extract recyclable materials with considerable energy and determination. They have little or no awareness of the health hazards associated with their activities. The major recycling materials recovered include paper products (cardboard, packages, newspapers/magazines and office waste paper), plastics, glass and metals, including aluminium.

There is no proper place for sorting these materials or storing them. Nor do the operators have appropriate equipment for handling their goods. For example, due to the lack of proper material recovery facilities and machines for compressing, baling and shredding, operators make do with improvised technologies based on their own ingenuity to handle waste materials. Generally these activities detract from the aesthetics of the environment and create an eyesore offensive to the minds and psyche of people who appreciate cleanliness. There is thus considerable chaos surrounding the ineffective, undefined, rudimentary and unregulated informal recycling systems. Clearly, for recycling to be effective, planned depots for recovered waste materials and material recovery facilities in line with the Abuja master plan are needed. The following classes of waste material collectors operate in Abuja (Chukwuocha 2004; EEESL 2007).

Waste Generators

At points of generation, some waste generators scavenge through their waste to redeem certain materials for direct re-use before storing the rest for collection or disposal.

Street Waste Material Collectors

These are groups of youths who typically sort through waste receptacles or open dumps along city streets in built-up areas before authorised waste-collection vehicles arrive. In the process, they litter the environment by scattering and tearing the garbage bags at street dumps or in storage areas, in order to pouring out the wastes for quick sorting. In most cases, they use handcarts or sacks over their shoulders to store and transport their recovered materials and a hook or curved iron for probing deep into heaps of waste.

Operational Crew Collectors

Waste collection crews of licensed collectors (government/private) undertake a certain amount of casual segregation when loading waste into collection trucks. The recovered items are stacked inside or on the sides of their vehicles, depending on type, for direct use or sale to dealers in waste materials. This resource recovery conducted during routine collections leads to loss of man hours.

Resident Waste Material Collectors

These waste collectors reside at or near the dumpsites. They live and work at these sites, at times with their families, under precarious weather conditions, ranging from extreme cold and wetness to great heat. After sorting and recovery, they stockpile recovered materials at the site according to type (plastics, paper, glass, etc.). Their numbers depend on the type and size of the dumps and the frequencies with which disposal vehicles visit the sites. When the stockpiles are large enough to make up a lorry or trailer load, or the available storage space becomes choked, they look for buyers (second-hand materials dealers), who are the middlemen who supply the manufacturers and exporters.

Trade-by-Barter Waste Material Collectors

Such collectors travel from house to house to barter new items made from waste materials, such as plastic bowls and buckets, for used or waste materials (e.g., different sizes and shapes of chemical, drug and beverage containers made of plastic, metal or glass). They are the agents of the middlemen. They use hand-drawn carts for this purpose. At times, when the material to be exchanged is heavier, or the owner wants an outright sale of his material, the collector prices it on behalf of his boss or may invite the boss to come to negotiate.

Middlemen/Second-hand Waste Material collectors, merchants ("pantaker") and dealers

These operators are the interface between recycling plants/exporters of recovered waste materials and local retailers and waste material collectors. Most members of this group are agents working for the cabals that own and sponsor the supply network. They procure materials from waste collectors at a token price, while themselves making huge profits without exposure to the hazards of scavenging. Their profits are not taxed.

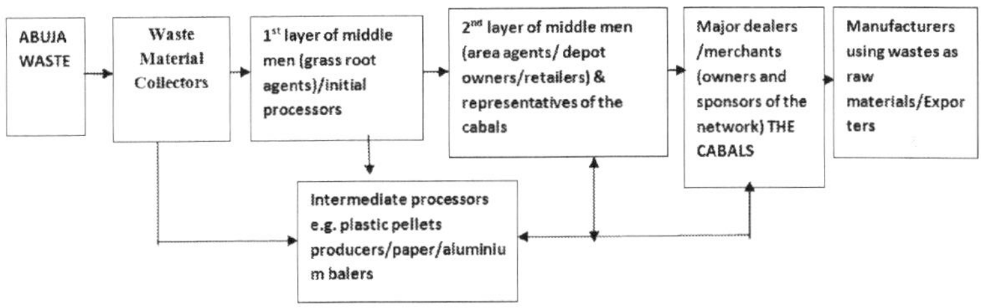

Figure 10.1:
Abuja waste-material recovery and supply network (*Source*: the Author 2013)

Unlike market distribution channels, with multiple options from which to choose and base strategies on, there is only one channel in this supply chain, with waste material collectors at the front end (see Figure 10.1). They are perpetually exploited by other members in the channel. There is a need for them to be equitably rewarded from the proceeds of the market.

Informal Sector Waste Recycling Challenges in Abuja

Dump Site Management

Waste material collectors conducting their activities at disposal sites have a tremendous influence on the discharge patterns at the sites. This influence runs counter to acceptable patterns of discharge. There is much indiscriminate dumping and littering at sites, which clutters available workspaces, reduces dump capacity and makes it difficult to maintain good standards of sanitation at sites.

Lack of comprehensive data/information on recyclable materials

No comprehensive data exist at government and informal levels for the proper planning, management and development of the recycling sector. For example, there is no information on the recyclable materials generated in, imported into or exported out of Abuja, the supply and demand situation, or the number of trips, location and capacity of the transport fleet over time. It was noted that most waste paper materials from Abuja end up in Jebba paper mills in Kwara state, Nigeria. However, there are no data on the quantities and qualities generated, recovered and transported to Jebba. The same is true of other materials in terms of quantities and qualities recovered, sold and supplied and the records of transaction at points of origins. Therefore, a comprehensive database is necessary to provide the information required for the sector's development.

Socioeconomic Impact

Collectors' earnings are below the poverty level or national minimum wage, as they sell their goods at give-away prices to middlemen, who exploit their ignorance and low literacy. The middlemen process and sell these materials at higher rates, and without tax, to manufacturers or exporters, who in turn make more profits. They are stigmatised and neglected by society as people of no social class and identity. Security agents and regulators constantly harass them and extort from them because they have no valued and valid social identity. It was also observed during key informant interviews that some of these collectors, especially those on the street, engage in light robbery, especially from unfenced premises in secluded areas. Items pilfered range from clothes drying outside to valuable items left unsecured in homes and offices. In some cases, homes are broken into, and collectors can be used for trailing or monitoring the movements of people for the purpose of such crimes as robbery, kidnap, assassination or terrorism. The average age of waste collectors in Abuja is between 15 and19 years. They have poor educational backgrounds, 80 per cent of them being elementary school dropouts (WEP 2004; EEESL 2007). They also lack management skills; capital for investment in transport, equipment and a legal business; and accommodation for themselves and their businesses. However, they are determined to survive against all the odds.

Health Impact

Most waste collectors are under the influence of drugs and many of them are disease vectors. Lacking adequate personal protective equipment and ignorant of other safety procedures, they are exposed to all sorts of health hazards. They live and work exposed to cold and heat and other diseases. Health hazards commonly experienced by waste pickers include poisoning from waste substances and inhalation of noxious gases; severe cuts from sharp objects; insect and reptile bites and stings; over-exposure to sun and rain; and strong stenches (WEP 2004; EEESL 2007). The most common diseases afflicting collectors are typhoid, diarrhoea, cholera and tetanus.

Handling of Recovered Waste Materials

Undeveloped plots and open spaces are accessed illegally and used as depots for recovered waste materials. This situation compromises the environmental aesthetics of the city and creates habitats for pests and disease vectors. Moreover, most recovered waste materials are, after being transported from original points of generation to final disposal sites some kilometres away, recovered and transported back to these illegal depots, which are situated almost at the points of origin. Similarly, when compliance drives catch up with these illegal depots, the owners either access new sites or have their wares transported back to the disposal sites where they originated.

Figure 10.2

Figure 10.3

Figure 10.4

Figures 10.2–10.4: Ingenuity in action: Laying cans on road to be compacted by passing vehicles (*Source:* the author)

Despite the rudimentary transport, bailing, compacting and packaging technologies available to it, the informal sector is not lacking in ingenuity, a quality it should be commended for. A further example of this ingenuity is the practice of laying beverage cans on the surfaces of remote roads, where vehicles flatten them as they pass, thereby compacting them for purposes of easier transportation and adding value to each load (see Figures 10.2, 10.3, 10.4).

Prospects for Informal Sector Waste Recycling in Abuja

Integrating Informal Waste Recycling into the Main Waste Management Stream

The informal sector should be mainstreamed into the waste management spectrum. For example, waste pickers can be designated as official used-materials merchants, given training and have their status upgraded by government. The existing activities and structures of the informal sector should be formalised and scaled-up to foster entrepreneurship that will support economic growth and integration of the sector into national, regional and global value chains. Formal recycling should reduce the costs of waste management and the amount of waste sent to final disposal sites, and contributes to the formal economy. Effective waste recycling reduces the carbon footprint of industries, thereby creating opportunities for economic gain from the global carbon trade (Roberts 2012).

Cooperative Societies and Stakeholder Partnerships

There is a need to open up opportunities in the informal sector for gainful employment that will improve the socioeconomic, health and housing status of the sector. By empowering people with training and tools, it will be possible to foster the emergence of small- and medium-scale entrepreneurs. Powerful groups of cooperative societies should be promoted in the informal sector, focused on the recycling of plastics, paper, spent oil, ferrous/non-ferrous metals, composting material, etc. Such societies will facilitate access to financial and technological aid and partnership ventures with stakeholders to support human resource development, skills acquisition and the provision of appropriate business and residential accommodation.

Security, Health, Safety and Environment

The author advocates registration of waste material collectors with local authorities and security agencies to secure operational permits and identification. This will reduce the security and socioeconomic challenges faced by the collectors. The need for personal protective equipment cannot be overemphasised, because safety is paramount, and if there is no safety, there is no business. Waste material collectors are also important to preventive medicine by promoting public and environmental health.

Technological Innovation

The recycling sector needs technological innovation, since manual sorting is labour intensive and carried out in unhygienic conditions. Examples of better technology include separators that sort materials by density, magnetic and chemical properties. Transport systems, equipment and tools are major constraints in the waste management chain, and need to be addressed. Research institutes, academia, financial institutions and the waste management industry need to partner on existing research that can be affordably commercialised, while new research can be undertaken on readapting existing local tools and equipment. Both initiatives could solve the technological challenges facing waste recycling and improve socioeconomic growth in the country.

Policy Review

Formal waste recycling is tied to the formulation of appropriate policies and legislation that can promote sustainable development. The socioeconomic, technological, health and safety issues facing the informal sector can only be changed through appropriate and comprehensive control measures. The time has come to review and update existing waste recycling laws and policies with a view to promoting wealth- and job-creation.

Pilot Programmes

There is a need for pilot waste recycling projects in six communities in Nigeria, one in each geopolitical zone. These would use ideal waste management practices and comprehensive master plans based on existing sociocultural and technological development levels, as illustrated by the Swedish waste management experience. Subsequently, the pilots, if successful, could be replicated in local government areas of Nigeria and adopted as a model for developing economies to address poverty alleviation, unemployment, protection and pollution prevention. By this means, the ignorance, illiteracy and exploitation of waste collectors could be curbed and achievement of the objectives of Millennium Development Goal 7 on environment and health promoted.

Management/Education/Training

Good management, awareness and educational campaigns and continuing skills acquisition and development are fundamentals for a sustainable business, no matter what the scale. Such projects apply to all branches of management: human resource management to ensure the appropriate skills; financial management, for the control of capital, budgeting and accounting, profit and loss analysis; marketing management to identify market opportunities and research supply channel development and performance strategies; and operational management, which deals with how to separate, grade, clean and compact materials and transport maximum payloads and add value to the entire business.

Conclusion

It is hoped that if the existing activities and structures of the informal waste recycling sector can be transformed and scaled-up to foster entrepreneurship, in conjunction with appropriate and comprehensive policies and legislation and equitable rewards for members of the supply chain, the sector will witness rapid economic growth and integration into national, regional and global value chains. Formalising the sector will reduce the carbon footprint of industries, create opportunities for economic gain from the global carbon trade and thus attract investors into the sector. Appropriate and comprehensive control measures are needed to improve socioeconomic, technological, health and safety outcomes in the sector, thereby alleviating poverty in the country and helping to achieve sustainable development goals, including Millennium Development Goal 7 on environmental sustainability.

References

Berger-Soge-Unecon (August 1982). Abuja Concept Plan for the Solid Waste Collection and Disposal System for the Federal Capital City, Abuja.

Berger-Soge-Unecon (August 1982). Abuja Concept Plan for the Liquid Waste Collection and Disposal System for the Federal Capital City, Abuja

Chukwuocha, E.O. (2003). Constraints of management of liquid waste. Liquid Waste Management Departmental Report, Abuja Environmental Protection Board.

Chukwuocha, E.O. (2004).Health and Social Implications of Scavenging. Paper presented at Women Environmental Programme (WEP) workshop on socioeconomic impact of scavenging in an emerging economy, 13March, Abuja.

Chukwuocha, E.O. (2010). Management of Guinea Current Ecosystem through Proper Waste Management and Development. Paper presented at international workshop: Private sector waste collection, disposal and recycling systems: The way forward on combating living resource depletion, pollution and degradation in the Guinea Current Large Marine Ecosystem through ecosystem-based regional actions, 19-21 October, Accra, Ghana. UNIDO GCLME project.

Emoc Engineering and Environmental Services Limited (EEESL) (2007). Survey of Private Sector Participation in Waste Management in the Federal Capital City Abuja.

Ezeah, C. (2012) Best Practice Drivers, Business Cases and Global Markets. Paper presented at the international workshop: S*ustainable waste management in Nigeria, 26-27 April, Abuja. Two-day CPD short course*, University of Wolverhampton.

Federal Republic of Nigeria (1979). The Master Plan for Abuja The New Federal Capital of Nigeria.

Nana, Y.A.I. (2010). The Collection, Disposal and Recycling Systems of Selected Private Sector Companies in Ghana. Paper presented at international workshop: Private sector waste collection, disposal and recycling systems: The way forward on combating living resource depletion, pollution and degradation in the Guinea Current Large Marine Ecosystem through ecosystem-based regional actions, 19-21 October, Accra, Ghana. UNIDO GCLME project.

Nzeadibe T.C., R.N.C. Anyadike and O. Adama (2012).Informal Recycling Patterns, Material Linkages and Urban Governance in Nigeria. Paper presented at international workshop: Survival in the Margins of African Cities: Informal Solid Waste Management in Nigeria, 21 August, Abuja.

Roberts, C. (2012). Waste to Wealth. Paper presented at the international workshop: S*ustainable waste management in Nigeria, 26-27 April, Abuja. Two-day CPD short course*, University of Wolverhampton.

Recycling & Waste World (2012). Mark Allengroup, Jessesferm, Snow Hill Dinton Salisbury Sp35HN

Waste Management (1999). The Swedish Experience. Stockholm: Gruphium Norstedts Tryckeri, pp.10-14.

Waste management World (2004). Review Issue 2004-2005; International Solid Waste Association: James & James London Uk

Women Environment Program (WEP) (2004). Study of Health and Social Implications of Scavenging in Dump/Material Transit Sites Abuja.

11

THE INFORMAL PRIVATE SECTOR IN INTEGRATED SOLID WASTE MANAGEMENT IN LAGOS

Olugbenga Olayinka Adebola

Introduction

Lagos is one of the 36 states in Nigeria. Although it is the commercial and industrial nerve centre of the country, it is the smallest in size (3,600 square kilometres) and has an estimated population of about 17.5 million. It presently has a very high population density of over 4,000 persons per square kilometre. According to UN estimates and assuming a 6 per cent growth rate, by 2015 Lagos will be the third largest mega-city in the world after Tokyo and Bombay/Mumbai (Government of Lagos State 2013). The state is presently divided into 20 local government and 37 local council development areas, a total of 57 municipalities. Under the present federal constitution of Nigeria, it is the responsibility of local (municipal) governments to manage the waste generated within their respective localities. However, Lagos's peculiar metropolitan status does not allow this constitutional framework to operate.

The realities and prospects of a population explosion, serious urbanisation and rapid industrial growth will have a direct effect on waste management systems in the state. The volume of waste being generated by this large population and the industrial sector (over 70 per cent of total industries in Nigeria are based in Lagos state) is enormous, overwhelming all government agencies involved in the collection, transportation and disposal of solid waste in the state. The inability of government agencies to cope adequately with the upsurge in municipal solid waste has led to a vacuum in the collection, transporting, recovery, recycling and disposal of solid waste. This vacuum has been filled by both the formal and informal private sector. The formal private sector is made up of registered companies, which usually have financial strength, some level of technical expertise and are generally mechanised. On the other hand, the informal private sector is made up of local cart builders (who construct carts in large numbers for rental), cart pushers (involved in house-to-house waste collection, transportation and recovery), waste pickers (involved in on-site waste recovery or waste picking), resource merchants (who purchase recovered recyclables and reusable waste) and recyclers (micro- and small-scale recyclers who convert recovered paper, aluminium, animal by-products, plastics and scrap metals into valuable materials and raw materials for consumption by the industrial sector).

The involvement of private sector participants in waste management in Lagos was, it must be stressed, the result of the irregular and erratic services of the government-owned agency responsible for waste management in the state. The nuisance caused by uncollected waste within communities led to the public themselves seeking an alternative waste disposal system, hence the entry of the private sector into solid waste management in Lagos state. In the early stages of private sector involvement, the neighbourhood structure determined the type of private sector collecting, transporting and disposal of the waste. Thus, the formal private sector operated in high-income, low population areas, while informal cart pushers and the like operated in low-income areas with a high population density. While the high-income neighbourhoods can afford formal, mechanised private sector operators, low-income neighbourhoods can only afford the hand cart pushers, whose operations and services are very inexpensive. However, with time that has changed, and both the formal and informal private sectors operate side by side now.

Adebola (2000) reported that in 1996-97, government in Lagos state started to recognise and formalise the private sector, especially the formal private sector, by introducing a pilot project on private sector participation in solid waste collection and disposal in Somolu and Kosofe local government areas. It aimed at test-running the capacities of the private sector and the willingness of the public to pay for the collection, transportation and disposal of their wastes, which hitherto had been a social service. The pilot scheme involved 22 private companies, including the author's, and marked the beginnings of private sector formalisation in Lagos's waste management. Unfortunately, government has done nothing to recognise the activities of the informal private sector, which is now a major stakeholder in solid waste management in Lagos. This chapter highlights the important roles of the latter sector in integrated solid waste management (ISSWM), especially waste collection, transportation, recovery and recycling. It also showcases the innovative aspects of ISSWM in job creation, poverty alleviation and reduction within the waste management sub-sector in the state. The unique features of the group and its interrelationship with other actors in the waste management industry are also explored. Most importantly, the group's role in promoting progress towards the achievement of certain Millennium Development Goals (MDGs) is highlighted. This chapter is a revised version of a paper written by the author on the operation of the informal solid waste management private sector in Lagos State in 2004.

Waste Management Structure in Lagos State

Local government

Waste management is the responsibility of the local governments in Nigeria in terms of Section 7, 4th Schedule of the federal constitution of Nigeria (Provision and maintenance of public convenience, sewage and refuse disposal). At the local government level, solid waste management is handled by waste/environmental monitoring units, with a chief environmental health officer (aka *Wole-Wole*) at the helm and under the direct supervision of the executive chairman or head of the local government. The environmental health officer is responsible for policy formulation, implementation, inspection and monitoring as well as prosecution of environmental offences at the local level.

Ministry of the Environment

The state government, through the Ministry of the Environment (MoE), oversees the overall cleanliness and sanitation of the state. The ministry has several departments (environmental services, drainage, signage and advertising, finance, waste monitoring, etc.) through which it carries out its statutory responsibilities. The ministry is responsible for overall policy formulation, implementation, monitoring and inspection and the prosecution of environmental offences across the state.

The ministry is also involved in policy implementation through its dedicated parastatals. These include Drain Ducks, responsible for clearing and maintaining the drainage network in the state and the Lagos State Waste Management Authority (LAWMA), which was initially responsible for general waste management (waste collection, transportation, storage, disposal and management of government-owned disposal facilities).

In the last few years, solid waste management has oscillated among state government agencies and local governments (Adebola 1998). Table 11.1 outlines these oscillations, which have even included the office of the state's deputy governor. These changes can be attributed to incessant political interference in the past in solid waste management functions. As expected, the changes result in different institutional solutions being brought to bear on waste management problems. This operational and institutional instability denied management agencies the benefit of accumulated experience and dampened staff morale.

Table 11.1: Institutional changes in State Solid Waste Management

Years	Name of institution	Supervising Authority
1977	Lagos State Refuse Disposal Board (LSRDB)	Ministry of Works and Transport
1980	Lagos State Waste Disposal Board (LSWDB)	Ministry of Works and Transport
1991	LAWMA	Ministry of the Environment and Physical Planning (MEPP)
1994	Local government councils and LAWMA	Local government and MEPP
1997	LAWMA and Private Sector Participation (PSP) pilot scheme in Somolu and Kosofe local government areas	Ministry of the Environment and Physical Planning (MEEP)
1998	PSP in domestic waste management state-wide and LAWMA handling industrial waste	MEEP/local governments
1999	PSP in waste management	All local governments/ MoE/office of the deputy governor
2004	Major PSP in waste management	MoE/office of the deputy governor
2007	Mega PSP in waste management	LAWMA/MoE
2009	Franchised mega PSP in waste management	LAWMA/MoE

Source: (the author 2009)

This institution was created around 1977 as the LSRDB with responsibility for collecting and disposing of refuse generated in the state. However, between 1977 and 1991, the agency underwent several changes in name and function before metamorphosing into what is now known as LAWMA. The change in name and status brought more responsibilities to the organisation, including collecting, transporting and disposing of all industrial and domestic waste, garden and agricultural waste, construction waste, clinical and commercial waste, etc. Above all, the organisation is responsible for managing all government-approved landfill in Nigeria sites in the state.

However, as Lagos became urbanised and industrialised and experienced a major population explosion, the volume of solid waste generated daily was so enormous the agency on its own was unable to cope. Consequently, it resorted to the provision of selective services, which became more erratic. A vacuum was thus created, giving an opportunity to the informal sector to fill the gaps.

Informal Private Sector

Informal private sector activities in solid waste management in Lagos state have been vital and highly significant to both the state and the country at large. The sector, which encompasses all aspects of integrated solid waste management (collection, transport, recovery, recycling and merchandising of recovered and recycled materials) is an institution in its own right. It represents a very effective operational network and its valuable products are distributed both within and outside the country. The actors in the informal sector and their role in solid waste management in Lagos are discussed below.

Cart pushers

These are involved in house-to-house waste collection for a fee, using specially built hand-pulled carts. They were brought into the industry as a result of the ineffectiveness of the government agencies responsible for collecting, transporting and disposing of waste. At one stage, it was estimated they were collecting over 70 per cent of the total waste generated in the state and transporting it to transfer bins strategically located by the government agency (Afon 2007). This group is also involved in waste recovery (the members go from house to house collecting waste at an agreed fee) and also sorts waste and recovers reusable and recyclable materials before disposing of the residue. Figure 1 shows a typical cart pusher moving along the street from house to house to collect and also recover waste. The non-formalisation of their activities makes it almost impossible to quantify the impact of the group within waste management with any accuracy. However, pushers cart away several hundred tons of waste per day and make a daily average of ₦ 2,600 (the equivalent of US$ 16.25 in 2006), whereas the average basic salary of government employees is only about ₦ 18,000 (US$ 112.50) per month. In Lagos state alone, it is estimated that over 5,000 cart pushers are operating within the informal waste industry.

Figure 11.2:
A typical cart pusher in Lagos (***Source: the author 2004)***

Cart builders

Here is another group of locals playing a major role in informal private-sector ISWM. Though very few in numbers, they invest their money in the construction of the carts used by the cart pushers (see Figure 11.2). The number of carts constructed and hired to cart pushers by individuals ranges from five to 50, depending on the financial resources of the cart builder. This business venture is very lucrative, with a high return on investment: each cart is hired out for between ₦ 250 ($ 1.56) and ₦ 500 ($ 3.13) per day, depending on area and locality, whereas the cost of constructing a standard cart is between ₦ 40, 000 and ₦ 50, 000 ($ 250.00 to $ 312.50). Within a few months of construction, each cart will have paid for itself and generated a reasonable profit with few overheads.

Waste pickers

This is the only group so far identified in Lagos state that is involved in both onsite and off-site waste recovery. Its members recover reusable and recyclable materials such as plastics, aluminium and other scrap metal, glass, paper and animal wastes (horns, bones, etc.). They were initially known as scavengers, but this term has become derogatory, especially considering the enormous value of their recovery of resources from waste. The name waste picker(s) became globally acceptable after the first world conference of waste pickers in 2008. It was also believed that adopting a single name would help in global identification and communication, because previously these operatives went under different names across the globe. They are also called resource recovery personnel. Some of them go door-to-door to recover reusable and recyclable materials, but most focus on the waste brought to disposal sites by the formal private sector and government agencies. In some cases, waste pickers also process part of the recovered waste before selling it either to resource merchants or directly to recycling industries. These processes include washing and burning (for instance, burning recovered electrical cables and wire to retrieve copper).

The majority of waste pickers live in shanties or makeshift houses built on or near disposal sites. It is common to see within such settlements heaps of recovered materials awaiting purchase or transport to recycling companies. Several million naira worth of materials (Odunaiya 2002) are recovered yearly by the over 4,000 informal waste pickers in Lagos state. The globalisation of demand for secondary raw materials has not gone unnoticed by the waste recovery and recycling system in Lagos. Only about 50 per cent of recovered items serve as raw materials to industries within Lagos state and the entire country, the remainder being exported to other countries in Africa (Ghana, Togo, Cameroon, Mali, Niger, Sudan, etc.) for both industrial and personal use.

Resource Merchants

These are the traders involved in purchasing all the recovered recyclable and reusable materials from waste pickers. They are also known as itinerant waste buyers. Some members of this group are ex-waste pickers, who now scout for materials at the sites either due to age or financial advancement. They are so wealthy that some are involved in the export of recovered resources to other countries, thereby earning foreign exchange. They are also influential enough to secure local purchase orders from companies to supply recovered waste materials.

Recyclers

This group includes both micro- and small-scale recycling companies, and converts recovered waste paper, aluminium, animal by-products, plastic, scrap metal, etc, into valuable materials and raw materials for consumption by the industrial sector. The recycling sector is a multimillion-naira enterprise, involving specialised equipment and machines to convert the recovered items into finished products or raw materials for use in several other applications. For instance, a company in Ikeja area of the state uses recovered waste paper to produce toilet tissue and exercise books, while another uses recovered plastic waste to produce plastic pellets for use in the production of new plastic products (hangers, toys, plastic hose, etc). Some of these recycled products and raw materials are exportable and generate foreign exchange. These recycling industries provide more than 2,000 jobs for different categories of the workforce.

Unique features of ISSWM in Lagos

- The operators cut across every known cultural and religious grouping in Nigeria, unlike in some cases where all the operatives in the informal waste management private sector share the same background. For example, ISSWM in Cairo is dominated by the Zabbaleen, a predominantly Christian group who migrated from the rural Assuit region in Southern Egypt.
- About 20 per cent of this group went into this sector to raise start-up capital for other businesses. Some even support themselves through secondary and tertiary school (Adebola 2003, 2004).

Achieving MDGs

Some of the contributions of the informal private sector towards the achievement of MDGs in Lagos (according to Adebola 2003, 2004) are listed below:

- **Employment opportunities:** There have been huge employment opportunities created by this sector in Lagos.
- **Foreign exchange earnings**: This sector has generated foreign exchange earnings from the export of recovered materials to other African countries.
- **Poverty reduction:** About 20 per cent of informal waste sector operatives entered the sector to raise start-up capital for other businesses.
- **Education:** A large percentage of this group pay their way through secondary and tertiary schools to enhance their quality in life by escaping poverty.
- **Access to healthcare:** Operatives in the sector can better afford medical bills in a country where it is extremely difficult to get free medical treatment.

The MDGs the informal sector in waste management in Lagos is moving closer to achieving are (UN 2008):

- **Goal 1: Eradicate extreme poverty and hunger** – The targets are a) halving the proportion of people living in extreme poverty (less than $1 /day) by 2015; b) achieving full and productive employment and decent work for all, including women and young people; And c) halving the proportion of people suffering from hunger by 2015. Through their activities, operators within the informal waste sector are helping to halve the proportion of people in extreme poverty, and clearly so, as most of them earn more than a dollar per day. They have also achieved full and productive employment (which is decent in its own right, except for the public perception) for both women and young people. Consequently, they have reduced the proportion of people in extreme hunger.

- **Goal 2: Achieve universal primary education** –The target is to ensure that by 2015 boys and girls everywhere will be able to complete primary schooling. The informal ISSWM operators are also helping to achieve this goal, as they now have enough muscle to pay for their wards' schooling (even though Lagos state government offers free education from primary to secondary school, there are still some financial commitments required from parents)
- **Goal 3: Promote gender equality and empower women** – The target is to eliminate gender disparities (in primary and secondary education). Beyond equality in education, the informal sector also empowers women economically and financially, thereby promoting gender equality.
- **Goal 7: Ensure environmental sustainability** – The targets are integrating the principles of sustainable development into country policies and programmes and reversing the loss of environmental resources. The activities of informal sector actors, most especially waste reuse, recovery and recycling, help to reverse the loss of environmental resources and reduce the generation of greenhouse gasses, which have an effect on climate change and global warming.

Conclusion

This chapter has shown that ISSWM is playing a major role in job creation, poverty reduction, environmental sustainability, and in promoting industrial activity in Lagos and national economic growth. Documentation of the activities of these informal private workers will be the first step towards the formalisation of their roles and status. Based on several advocacy programmes, the government is planning to formalise the activities of the informal sector and afford it much needed recognition. Operatives have also formed themselves into groups and an association to cater for members' welfare. They also have established a sound relationship with other major actors in Lagos's waste management industry.

References

Adebola, O.O.(1998) Private Sector Participation (PSP): Alternative waste management system for Lagos state. Paper presented at the State of the Environment lecture series, organised by Waste Pilot Scheme Operators Group.

Adebola, O.O.(2000). Sustaining Private Sector Participation (PSP) in solid waste management in Lagos, Nigeria - A developing country. Paper presented and published in the final CD ROM of the 5th World Congress and Envirotech Trade Show in Toronto, Canada.

Adebola, O.O. (2003). Management of a landfill in a developing nation: The effect of the scavengers on onsite operations. Proposal submitted to Lagos State Ministry of the Environment.

Adebola, O.O. (2004). Independent study on the operations of the informal private sector in solid waste management in Lagos state. Lagos: Richbol Environmental Services.

Afon, A.O. (2007). Informal sector initiative in the primary sub-system of urban solid waste management in Lagos, Nigeria. Habitat International 31 (2): 193 – 04.

Government of Lagos State (2013).Population of Lagos state [Online]. Retrieved 20 September 2013 from http://www.lagosstate.gov.ng/pagelinks.php?p=6

Iskander-Kamel, L.R (1994). Mokattam Garbage Village, Cairo, Egypt.and notes from a Lecture delivered by Dr. L. Iskander on 16th April, 2002 at Cairo, during a study tour organised by DIN and sponsored by Ford Foundation - the author was also part of the tour.

Odunaiya, C.O.(2002). Repositioning solid waste management practices in cosmopolitan Lagos.Sixth Inaugural Lecture Series, Lagos State Polytechnic.

United Nations (2008).Official list of MDGs indicators. Retrieved on 20 September 2013 from http://mdgs.un.org/unsd/mdg/host.aspx?Content=indicators/officiallist.htm

12

QUESTIONING THE ROLE OF INTERNATIONAL ORGANISATIONS IN SOLID WASTE MANAGEMENT IN ADDIS ABABA

Camilla Bjerkli

Introduction

International organisations have been paying increased attention to improving urban services such as solid waste management. There is agreement that waste management in many African urban areas is in crisis and that technical, institutional and financial investments are needed to improve it (World Bank 2004). Due to concerns over the effectiveness of aid, "good governance" has become the recipe for improving public services. The official belief is that improving the capacity of governmental institutions in partnership with non-state actors will lead to more efficient implementation and management of public services (UN Habitat 2010b; World Bank 2008; World Bank 2004; UN Habitat 2010a). This has resulted in increased direct budget support to strengthen recipient countries' policymaking and management systems and institutional capacity (World Bank 1992; Bulkeley et al. 2003; UN Habitat 2002a; Smith 2007; World Bank 2008; Allen et al. 2008; UN Habitat 2002b,; African Development Bank 2002; Hinton and Groves 2004). It is recognised that by bypassing the state and entering into direct partnerships with non-state actors, government's ability to manage public services can be reduced. This may be a reasonable assumption, but what if the recipient government institution is incapable of providing honest service delivery? Direct budget support to governments is justified on the dubious assumption that governments are representative and accountable to their citizens. This also means that it is up to the government to decide who is to receive support, what kind of support and who is seen as a partner.

The implementation of projects to support governments in improving public services in line with "good governance" has in practice been more problematic than in theory (Grindle 2010; Jean-Pierre and Sardan 2008; Groves 2004; Hinton and Groves 2004; Myers 2005). Many have argued that projects are affected by power relations, actors, agendas and the local context in which projects are implemented (Crespin 2006; Jean-Pierre and Sardan 2008; Pierre 2000; Hoebink 2006; Grindle 2007; Andrews 2008; Santiso 2001; Irish Aid 2008; Chambers and Pettit 2004). Studies have shown that governments have adopted and implemented projects and policies, but that few positive results are seen on the ground. One can say that projects are mostly used in a rhetorical way, resulting in a gap between what is reported and what is done (Schalkwijk 2005; Jean-Pierre and Sardan 2008). Ferguson (1994) shows how an agricultural project in Lesotho was used by government to gain political control over the opposition and strengthen its own power. Ferguson (1994: 255) concludes:

> ...the "development" apparatus in Lesotho is not a machine for eliminating poverty ... it is a machine for reinforcing and expanding the exercise of bureaucratic state power, which incidentally takes "poverty" as it point of entry – launching an intervention that may have no effect on the poverty but does ... have other concrete effects.

Increased acknowledgement of the rhetorical use of projects has led to an increased focus on power and politics in order to critically examine the way projects are carried out.

Studies also show that many donors ignore and fail to deal with the cultural and political aspects of their projects. Eyben (2004) and Groves (2004) argue that international organisations in many cases ignore the history of the country and fail to understand the culture and how power operates. Eyben (2004) further argues that donors tend to interact with those they feel most comfortable with, which in most cases is the elite. This provides donors with only a partial perspective on the situation and leads to empowering the elite while disempowering the poor. If international organisations wish to make a real difference, they have to focus more on issues of power and politics and the local context in order to understand the culture, politics and leadership.

This chapter focuses on how governance operates in Addis Ababa and how the involvement of four international organisations has affected the management of solid waste. I start by examining how the shift from "government" to "governance" has resulted in a rescaling and a shift in the power relations and actors involved in urban services.

Governance and the rescaling of power

There is agreement that more emphasis should be placed on critical examination of how governance processes operate at the local level. This is essential to understanding why and how projects aiming to improve urban services are adopted and implemented and the outcome of these processes (Jean-Pierre and Sardan 2008; Irish Aid 2008; Hinton and Groves 2004). Beneath the apparently strict descriptions of how practice is organised lies a complex network of actors and agencies that influence, form and reform decisions and practices. A more critical approach is therefore needed to understand the messy reality of interests and agendas behind the decisions taken on urban service provision. Further, local agendas and actions are part of a much larger network, where power exercised at local scale is influenced or enabled by agendas and strategies at higher scales (Dahlman 2010). We are therefore confronted with a complex network of actors, agendas and power relationships across scales

(Swyngedouw 2000).

The shift from "government" to "governance" in development discourse implies decentralisation and the involvement of non-state actors in policymaking and the management of urban services. Consequently, many cities have undergone decentralisation to give more power to lower government levels and to increase non-state actor participation in managing the state in a more "democratic" way (Swyngedouw 2005; Smith 2007). In urban management, these new horizontal structures range from actors operating on the local scale (often associated with non-state actors and local government), the national scale (often associated with the state) and the global scale (often associated with international organisations). According to Swyngedouw (2005), these new structures and networks have resulted in a rescaling of governance processes, which has resulted in a shift in power relationships among the actors involved in these processes.

In a theoretical perspective, the consequence of this restructuring is a weakened management role for the state. Non-state actors have in theory been given increased power to affect decisions. However, even though the structure and network of relationships has changed, this does not necessarily reduce the role or power of the state. How these governance processes operate depends on the political culture, the way power is distributed, which actor is seen as a partner and the nature of the relationships. According to Swyngedouw (2000), these power relations empower those actors who are seen as partners, but also exclude others from governance processes.

Smith (1992: 341) argues that the practice of power relations between state and non-state actors has resulted in the creation of new scales and what he refers to as the "politics of scales." The focus should therefore be on these power relationships and how conflicting interests and political struggles operate and create new scales for negotiations. These power relations across scales can change the role of a certain geographical scale and create new scales or arenas for negotiation (Swyngedouw 2000; Ekers and Loftus 2008). The politics of scale is relevant in showing how the involvement of international organisations in the governance process affects power relations between actors operating at the local scale.

Power is everywhere in society and to study power relationships it is useful to identify the types of power used by actors. Power is here seen as capacity, medium and technologies (Allen 2003). As capacity, power is seen as a property that can be held and delegated or distributed. Power is thus the ability to make decisions and to control how things are done or organised. Some actors have this power, others not. Some actors are therefore able to decide to make other actors (without power) do something they do not want to do. The state is a classic example of power as capacity: it has legitimacy and power to make laws and regulations that citizens have to obey.

Power as a medium refers to power in terms of resources, such as economic value, knowledge or collective actions. Power is therefore seen as something generated by the use of resources and refers to a range of social practices actors adopt to achieve certain goals or to react to existing power structures. Power as technology views power as relationships, which consist of a series of "techniques" of practice. It draws on Foucault's work on the circulation of power and how power is exercised and practised within a given society (Foucault 1982). According to Foucault, people regulate their own behaviour according to the norms of society. Power relations have therefore to be seen in relation to the culture: how people, through their acceptance of how they should act within a given context, affect how power is distributed and acted upon by the actors. Power is everywhere and is a series of practices undertaken by actors in their interactions with others (Gallagher 2008). According to Foucault, power is always calculated to satisfy various aims and objectives. This does not mean that the outcome of power relations is in line with the aims or objectives of an action. The complexity of the social world means that social actions always have unintended effects, since there are various options available for actors to act upon in power relationships. The intentions behind actions are therefore irrelevant: it is the nature and outcome of these power relations that are interesting (Foucault 1982).

For this chapter, the three approaches to power have been combined. Power as capacity and medium is included to be able to identify the various "tactics" used by actors in solid waste management, and power as technology is included as an approach to understanding how power operates within a given context.

Methodology

The empirical data for this chapter were collected during ethnographic fieldwork in Addis Ababa between 2009 and 2011. Fifty-four semi-structured interviews were conducted with city officials (14), international organisations (four), local NGOs (two) and actors in the informal recycling sector (34). My aim was to collect the subjective views of participants and explore their interests and agendas. It has thus been possible to identify conflicting interests and to better understand the underlying reasons for the actions taken. In most cases, follow-up interviews were conducted in order to track developments over time and the implementation and outcome of projects. Data were also collected through personal observation, participation in workshops and reviews of newspaper and official documents. This allowed me to explore official views and statements and to compare them with my own observations and the information collected from interviews. Fieldwork data were analysed using Constant Comparative Methodology (CCM) whereby interviews are compared, both within the same group of actors and between groups (Boeije 2002; Postholm 2005). This was essential for seeing if the subjective views of actors within and between groups were contradictory. Further, the data collected through interviews, observation and newspaper reviews have been related to context – to the actor's background and the relationships between me as researcher and participants, which may have been affected by power relations, interests and agendas (Silverman 2009; Gibbs 2007). CCM has also been useful for cross-checking information given to me by actors, in order to increase the reliability of the data collected.

The Ethiopian government: its nature and relations with international donors

Ethiopia is one of the largest recipients of foreign development aid in Africa and aid has played a major role in its development (Gebregziabher 2006). Particularly with the adoption of the Structural Adjustment Programme in 1992, the change towards a federal and democratic regime in 1995 and the Sustainable Development Programme for Reduction in Poverty (SDPRP) in 2002, the country has benefited from a significant amount of aid. According to OECD-DAC statistics, net Official Development Assistance (ODA) to Ethiopia was US$ 1.94 billion in 2006, making Ethiopia the seventh largest recipient among 169 aid-receiving countries. Since the second half of the 1990s, the amount of ODA has increased from an average of US$ 881 million per annum to an average of US$ 1,683 million per year between 2000 and 2006 (Geberegziabher 2006).

However, the national election of 2005 marked a change in Ethiopia's relationships with international donors. Opposition supporters protested of irregularities in vote counting and the Ethiopian government mounted a vicious crackdown, which resulted in 200 deaths and over 30,000 detentions (Tronvoll 2011). Many donors feared the Ethiopian government would use international financial support to secure its own power and decided to stop direct budget support to it. The World Bank, among others, committed itself to reducing aid over time if governance did not improve (Human Rights Watch 2010; World Bank 2004). However, in 2008 the World Bank and other donors claimed that Ethiopia had made progress in governance and money again flowed into the country to improve public services (Africa Development Bank 2009; Human Rights Watch 2010). It seemed that donors were ignoring their earlier concerns, and were instead viewing the government's undemocratic character as a technical challenge rather than as one of political will (Human Rights Watch 2010).

In theory, donors have power to influence policies in recipient countries to some degree, particularly where aid dependence is high. In Ethiopia, despite its poverty and aid dependence, donors seem to have only limited policy influence (Borchgrevink 2008). The Ethiopian government has used donor-supported programmes as a political weapon to control the population, undermine political opponents and secure its own interests and power (Human Rights Watch 2010). In addition, some recent policies indicate that the government is seeking to reduce the power of civil society and international donors. In 2009, it introduced the *Proclamation to Provide for the Registration and Regulation of Charities and Societies*(Federal Democratic Republic of Ethiopia 2009), which restricts local NGOs receiving more than 10 per cent of their financing from foreign sources in engaging in human rights and advocacy activities. The argument behind this law is that Ethiopia, if it is to have a democratic state with respect for human and civil rights, has to achieve this on its own and not through foreign donor funding.

In 2011, the government proposed a new regulation to the effect that no international projects would be approved if administrative costs exceeded 30 per cent of total costs. In theory this is good, but the vague definition of administrative costs meant that up to 70 per cent of the activities of certain international NGOs could be so designated. An interview with one such NGO in 2011 revealed that this regulation would make it even more difficult for it to continue working in the country. These new laws indicate that the government is turning donations away from civil society and into direct budget support to government and its agencies. In this political context, in combination with the international focus on "good governance," most donations now go through the government, with the official aim of improving governance and the management of public services.

Political culture within the city administration

Ethiopia has a relatively long history as a highly centralised state with upward accountability to traditional rulers and later the ruling party (Harold 1994; Pausewang et al. 2002; Kapuscinski 2006). However, Ethiopia is one of the few African countries to ratify the African Union's good governance proclamation and, in line with good governance principles, to have decentralised the state.The Ethiopian government has implemented two national decentralisation programmes and several policies, such as Business Process Reengineering (BPR), to improve governance and the efficiency of public services (Taye and Tegegne 2007; Human Rights Watch 2010; African Development Bank 2009. Despite these reforms and structural changes, many claim that the government is still highly centralised (Taye and Tegegne 2007; Aalen and Tronvoll 2009; Vaughan and Tronvoll 2002; Lefort 2010; Africa Development Bank 2009; Poluha 2004). To understand the political culture of Addis Ababa, a deeper understanding is necessary of the distribution of power within the society, based on the historical interactions of a social collective over time (Vaughan and Tronvoll 2002). Vaughan and Tronvoll (2002) describe Ethiopian social organisation as highly hierarchical, with socio-political positions and status clearly defined and widely understood within society. Men are superior to women; elders to the younger; and religious leaders and politicians have added authority. This culture and behaviour have been transmitted through generations and still affect socialisation in the country, with individuals ranked according to set criteria. This has resulted in a culture where people obey the orders of those of higher social rank and where officials in the city administration are rarely questioned by citizens. Similarly, these officials rarely question the orders they receive from above (Poluha 2004).

The hierarchical political and social organisation has meant that some groups of people are excluded from, and some are included in, political processes. People are mainly employed on the basis of party membership and loyalty, and much less because of experience, expertise or competency. This was confirmed repeatedly in discussions with participants (even so-called loyal participants) during fieldwork. Moreover, to ensure the loyalty of city employees at all levels at any time, the ruling party undertakes frequent evaluations of them. Those who are not loyal to the ruling party and its interests are fired and replaced by people who are. This has resulted in a high turnover of officials, with detrimental consequences for administrative continuity and institutional memory (Bjerkli 2012). Political control is further strengthened by the dual responsibility of the elected body in the city administration to formulate policies and plans and implement these decisions. Politicians are able to permeate various aspects of the city's responsibilities, such as licensing, bidding and contracts (Yirgalem 2008). This, in combination with upward accountability, has led to increased corruption and abuse of authority, and has further eroded public trust in the municipality (Meheret 1999; Taye and Tegegne 2007). On top of all of this, official salaries are low. To compensate, most officials spend much of their time applying for international projects, scholarships, and in arranging workshops and study tours. The net effect of all these factors is that limited attention is given to citizens and needs on the ground.

Aims and interests of international organisations in solid waste in Addis Ababa

The solid waste situation in Addis Ababa is in crisis, as in many other African cities (SBPDA 2003; Fikade 2009; ORAAMP 2002). The city has grown rapidly, and both its population and its size have more than doubled in the last 30 years, as has the waste generated in the city (UN Habitat 2008). The solid waste service today is poor and characterised by a low collection rate, high transport costs and the lack of a proper sanitary landfill (Fikade 2009). Furthermore, the service provided represents an "end of pipe" approach to solid waste, focusing only on collection, transport and disposal. Parallel to the municipal system is the informal recycling sector, which has been operating in the city for the last 80 years. This sector has developed a highly organised system for collecting, trading in and reusing solid wastes. Significant volumes of materials are involved and these activities are an important source of livelihood for many urban citizens (Bjerkli 2005; ILO 2009). Today, the informal recycling system mainly comprises five actors: the scavengers operating at the municipal dump site (approx. 500), iterant junk buyers (approx. 3,000), wholesalers (approx. 500), small-scale craftsmen and various industries (Bjerkill 2005).

Many international organisations have been involved in solid waste management in Addis Ababa over the last 10 years. Here, the focus is on four of them: the International Labour Organisation (ILO); the Income Generation and Climate Protection through the Sustainable Valorisation of Municipal Solid Wastes in Emerging Mega-cities (IGNIS), a research project funded by the German Federal Ministry of Education and Research; the French Development Agency (FDA); and the Public Private Infrastructure Advisory Facility (PPIAF), financed by 17 multilateral and bilateral donors. All these organisations aim to improve solid waste management in the city. ILO and IGNIS have been involved in projects and activities related to research, pilot projects and workshops. Their aim has been to improve solid waste services by focusing on reuse and recycling and improving collection and disposal of wastes through job creation, especially for marginalised groups (ILO 2007, 2008a, 2008b, 2008c, 2008d; Gebreselassie 2007; IGNIS 2009; Enda Ethiopia

2009). PPIAF has been supporting the city administration in its efforts to involve private sector partners in solid waste collection. The main focus has been on providing legal and institutional capacity to the administration for this purpose (PPIAF 2011). FDA has been supporting the city administration with financial and technical assistance to close the existing landfill and to construct a new sanitary landfill and four new transfer stations in the city. Table 12.1 gives an overview of these international organisations and their solid waste projects in Addis.

Table 12.1:
Aims and activities of international partners in Addis Ababa

Organisation	Aim	Activities	Partners	Funding
ILO	Job creation, health of waste workers, awareness	Research, pilot projects, workshops	Addis Ababa city administration (AACA) Enda Ethiopia, Environmental Protection Authority (EPA)	?
IGNIS	Job creation, health of waste workers, reuse and recycling	Research, pilot projects, workshops	AAU (Addis Ababa University), Enda Ethiopia, EPA	€ 4.38 m
PPIAF	Job creation and establishment of public private partnerships (PPP)	Technical assistance, capacity building	AACA	?
FDA	Improved collection and disposal of solid waste	Technical assistance, capacity building	AACA, Enda-Ethiopia, EPA	€ 20 m

Source: Author's fieldwork

Nature of international solid waste projects in Addis Ababa

As argued above, the implementation of projects has to be seen in relation to local context, including the actors involved, their interests and agendas and the power relationships between them. Generally speaking, there are five main factors affecting the projects carried out in Addis.

First, the government has been unwilling to loosen its control over civil society and has used its power to restrict the activities of both civil society and international organisations. The projects that have been accepted relate only to the technical and financial aspects of solid waste, not issues of politics and human rights. In most cases, international organisations are forced to enter into a partnership with the city administration in order to execute their projects. This has made it difficult for local actors to become partners in these projects. This is clearly visible in Table 1, where the international organisations have entered into partnerships with the city administration, the university and the EPA. The administration has used its power – in the sense of power as capacity – to introduce new laws that force international organisations to enter into partnerships with governmental agencies to the exclusion of other actors and potential partners.

Second, all the goals of and visions for the projects focus on principles consistent with "good governance," for instance PPP and integrated solid waste management (ISSWM). Very few if any local actors (such as pre-collectors and those involved in informal recycling) were given a chance to express their needs and interests when these projects were being developed. Moreover, in their efforts to become partners in international projects, local NGOs use most of their time writing project proposals and trying to meet the requirements set by international organisations, rather than focusing on the needs of actors on the ground. During my fieldwork, I observed a meeting where the donor listed the issues be included in a proposal focused on empowering women, job creation and the urban physical environment. Here, as in most other cases, when project goals and activities are defined, the needs of local actors were neither heard nor considered. Another example is an IGNIS project to create a new bag for the *korales* (itinerant junk buyers) to carry their items and a manual on how to carry and lift the bags (IGNIS 2009). Yet the *korales* did not see this as a priority: they were more concerned that the city administration was ignoring them in their plans to implement a modern, new solid waste system that had no place for them. This can also be identified as power as capacity, were local actors have few possibilities to express their needs and where the city administration and international organisations use their power to decide how things should be organised.

Third, as noted earlier, the culture within city administration is characterised by loyalty to the ruling party. This has resulted in upward accountability and a focus on acting according to party interests, with limited accountability towards local actors. This, combined with low salaries, means that city officials spend most of their working time being loyal, following orders and seeking to participate in projects supported by international organisations. Here, power can be viewed as both capacity and as medium, with civil servants acting according to ruling party interests and simultaneously using their positions to improve their income. For civil servants, international projects are seen as additional income and as possible access to study tours or training abroad (Søreide et al. 2012). During my fieldwork, it became clear that some local partners were participating in several international projects. Enda Ethiopia, AACA and EPA were partners of IGNIS, FDA and PPIAF. However, the international organisations were not made aware of other ongoing projects. This might indicate that it is not in the interest of local partners to provide this information, because decreased opportunities for extra income might result. Consequently, there has been no cooperation between projects and in some cases projects have been focused on the same activity, as, for example, the ILO, IGNIS and PPIAF activities to improve solid waste collection and the occupational health of pre-collectors. Here, power can be viewed as a medium, with civil servants trying to prevent cooperation between projects.

Fourth, in line with what Groves (2004) has argued, international organisations often fail to understand the local culture and to examine realities on the ground. As regards the projects identified above, all of the four international organisations failed to examine the needs of local actors and tended to ignore the political context of their projects. All of them seemed to listen unquestioningly to their partners and were not critical of the facts presented to them. FDA's involvement with the city administration in planning a new system for reuse and recycling is a good example of this behaviour. The city's long-standing informal recycling sector, an important livelihood source for several thousand citizens, was ignored (Bjerkli 2012). Further, to close the current landfill, FDA and the city hired a Finnish company. A Finnish foreign ministry press release announced:

> Good news from Africa: Finnish EKOKEM exports environmental expertise to Ethiopia. Climate sustainability, environmental protection and sustainable development are themes that Finland stresses in its development policy. Finnish Ekokem provides a practical example for how Finnish environmental expertise is exported abroad. The company's subsidiary Ekokem-Palvelu Oy is currently partially closing a landfill in Addis Ababa. Representatives of Ekokem stress that in addition to finding durable solutions for waste disposal as the population is growing, the project functions as a means to fight climate change. (Karlson 2012)

This is good news in terms of technical improvements, such as the long awaited closure of the old dump and establishment of a methane-capture plant. However, the release ignores the social aspects and the 500 scavengers who will lose their livelihoods. Furthermore, PPIAF, in supporting the city administration in creating PPPs, did not question the way in which this was done. According to PPIAF, their project led to 524 new enterprises and the employment of 5,815 pre-collection operators in the city. Consequently, the municipal garbage collection rate has risen from 60 to 80 per cent (PPIAF 2011). The accuracy of these data is questionable. The last waste generation study was done by Norconsult in 1982, since when the city has doubled in size and population and socioeconomic conditions have changed. This seems to be another case of uncritical acceptance of the information presented. Moreover, PPIAF's reading is very different from Zelalem's (2006) and Bjerkli's (2012). These authors argue that the private sector has been used as a political tool to create employment prior to elections. This was done by excluding existing actors in the city and hiring new employees loyal to the ruling party. Moreover, they suggest that there has been a reduction in the amount of waste collected in the city.

As Grove and Eyben (2004) point out, lack of critical sense about official data presented leads to an inadequate understanding of the reality of the situation. Moreover, this has also contributed to further exclusion of local actors involved in solid waste and to a strengthening of the power of the municipality to gain a "modern" high-tech solid waste management system. Here, we are considering power as technology, with international organisations interested in selling their own technology, even though it is unsuitable to the local context, and choosing not to act or be critical in order to achieve this end.

Fifth, the frequent evaluations based on loyalty and the high turnover among civil servants also affects the projects. In the field, I observed that civil servants sent for training abroad left their positions upon their return, either for better job opportunities in international NGOs and the private sector, or for political reasons, as they were either promoted or fired. This means that the training and capacity building offered to local partners has a very limited effect on local institutions. Moreover, the projects constantly have to deal with new people. In most cases, these new officials know little about the projects or about solid waste, largely because they are employed for their political affiliation. Additionally, the political culture of the city administration leaves officials reluctant to take decisions without clearance from those higher up, in many cases leading to delays in the projects.

The high turnover has also resulted in a lack of institutional memory, so that projects or studies tend to be forgotten or disappear. An example of this is the studies conducted by the ILO (ILO 2008a; ILO 2008b; ILO 2007; ILO 2009; ILO 2008d), of whose existence newly appointed officials were unaware. These studies were therefore not taken into consideration when new solid waste management plans were made. This can be seen as the power of technologies, with power as capacity as a side effect: the political culture and control by employers has led to a city administration lacking in institutional memory and capacity to implement the intended improved solid waste service.

Reinforcing government and excluding the poor

This chapter has tried to show how political context and power relationships among the actors involved have affected solid waste projects in Addis Ababa. As Swyngedouw (2000, 2005) argues, power relationships between actors at various scales create new arenas for negotiation. In Addis, a new scale of negotiation has emerged, in that the city administration and international organisations are included and have been able to express their interests, while local actors and intended beneficiaries have been excluded from expressing their views and taking part in projects. In line with Swyngedouw's (2000: 208) argument, the rescaling of the governance processes in Addis has not led to a reduction of state power but to a reinforcement of authoritarian power structures. The Ethiopian state and its relationship with international organisations seem to work as a closed circle, where resources and knowledge (salaries, official cars, workshops, scholarship, travel, etc.) circulate between the Ethiopian state and the organisations and rarely materialise on the ground. Local actors interact to a limited degree with the city administration and international organisations, through research and pilot projects, but they are unable to affect the projects due to the political context and their lack of power to act on their interests.

From the Ethiopian government's perspective, these projects are important in several ways. First, the government is interested in Western technology and high-tech systems, which represent development and modernisation. Second, the government may be using international projects as a way of justifying themselves and pretending they are doing something (e.g., creating 5,815 jobs), thus hiding the real situation on the ground, which has a life of its own. For civil servants, these projects hold out the prospect of increased income and it may also be that government uses international projects to secure civil servants' loyalty by placing them in positions from which personal benefits may be expected.

Further, project evaluations seem to be far more positive than what is observed in the city (Crespin 2006; Chambers and Pettit 2004). Based on their failure to be critical of project implementation and outcomes, one may question whether the real agenda of the international organisations is to empower the poor. Promoting governance without taking critical account of how the governance process operates in the local context means that projects deliver the opposite of what they promise. Today, it can seem that the ultimate goal of international projects is to finance the running of the city administration and the well-being of civil servants in order to get the administrative "machinery" to work.

By ignoring the political aspects of the projects, international organisations are continuing to reproduce the same international discourse and the same "white elephant" errors. Are they afraid of examining the political aspects of projects for fear or having to face the realities on the ground and the consequences of their involvement? Today, international organisations are acting on a scale that seems comfortable for them, without getting into politics and trouble. By so doing, they have become integral parts of existing, complex patterns of power in Addis Ababa. Without being too critical, international organisations seem to contribute to the reproduction and reinforcement of the prevailing patterns of exclusion and inclusion that they are trying to eliminate with their projects. A radical change is needed to strengthen the outcomes of projects and to make them truer to their own principles. In line with Ferguson (1994), this chapter has shown the need to critically examine power relationships, politics and the way governance operates in the local context; to critically examine project implementation; and, if needed, to withdraw from projects. International organisations need to move out of their comfort zone and to dare to require more from governments to ensure that their projects end up as more than a soon-forgotten paper exercise and result in improved development on the ground that will benefit the poor.

References

Aalen, L. and K. Tronvoll (2009). The End of Democracy? Curtailing Political and Civil Rights in Ethiopia. *Review of African Political Economy* **36**(14): 193-207..

African Development Bank (2002). Study on Solid Waste Management Options for Africa. Abidjan:ADB Sustainable Development and Poverty Reduction Unit. [

African Development Bank (2009). *Federal Democratic Republic of Ethiopia. Country Governance Profile.* Available at http://www.afdb.org/fileadmin/uploads/afdb/Documents/Project-and-Operations/Ethiopia%20%E2%80%93%20Country%20Governance%20Profile%20EN.pdf. Accessed 4 February 2012.

Allen, A., P. Hofmann and H. Griffiths (2008). *Moving down the ladder: governance and sanitation that works for the urban poor*. Symposium: Sanitation for the urban poor. Partnerships and governance. Netherlands: IRC.

Allen, J. (2003). *Lost Geographies of Power.* Oxford: Blackwell.

Andrews, M. (2008). The Good Governance Agenda: Beyond Indicators without Theory. *Oxford Development Studies* **36:** 379 – 407.

Bjerkli, C.L. (2005). The Cycle of Plastic Waste. An Analysis of the Informal Plastic Recovery System in Addis Ababa, Ethiopia. Master's thesis, Norwegian University of Science and Technology.

Bjerkli C.L. (2012). Governance on the ground: A study of solid waste management in Addis Ababa, Ethiopia. *International Journal of Urban and Regional Research* **37** (4):1273-1287.

Boeije, H. (2002). A Purposeful Approach to the Constant Comparative Methods in the Analysis of Qualitative Interviews. *Quality and Quantity* **36:** 391 – 409.

Borchgrevink, A. (2008). Limits to Donor Influence: Ethiopia, Aid and Conditionality. *Forum for Development Studies* **35** (2): 195 – 220.

Bulkeley, H., A. Davies, B. Evans, D. Gribbs, K. Kern and K. Theobald (2003). Environmental governance and transnational local authority networks in Europe. *Journal of Environmental Politics and Planning* **5** (3): 235-254..

Chambers, R. and J. Pettit (2004). Shifting Power to Make a Difference. In *Inclusive Aid. Changing Power and Relationships in International Development,* L. Groves and R. Hinton (eds). London: Earthscan.

Crespin, J. (2006). Aiding local action: The constraints faced by donor agencies in supporting effective, pro-poor initiatives on the ground. *Environment and Urbanization* **18:** 433 – 49.

Dahlman, C.T. (2010). Scale. In *Key Concepts in Political Geography,* C. Gallaher, C.T. Dahlman, M. Gilmartin, A. Mountz. and P. Shirlow (eds). London and California: Sage.

Ekers, M. and A. Loftus (2008). The power of water: Developing dialogues between Foucault and Gramsci. *Environment and Planning: Society and Space* **26:** 698 – 718.

Enda Ethiopia (2009). Project proposal on income generation and climate protection by valorizing municipal solid wastes in a sustainable way in an emerging mega-city, Addis Ababa. IGNIS.

Eyben, R. (2004). Who Owns a Poverty Reduction Strategy? A Case Study of Power, Instruments and Relationships in Bolivia. In *Inclusive Aid. Changing Power and Relationships in International Development,* L. Groves and R. Hinton (eds). London: Earthscan.

Federal Democratic Republic of Ethiopia (2009). Proclamation 12 for the Registration and Regulation of Charities and Societies.

Ferguson, J. (1994). *The Anti-Politics Machine. "Development," Depoliticization, and Bureaucratic Power in Lesotho.* Minneapolis: University of Minnesota Press.

Fikade, T. (2009). Situation Analysis and City Wide Strategy for Integrated and Sustainable Solid Waste Management in Addis Ababa. Addis Ababa: International Labour Organization/Addis Ababa City Administration.

Foucault, M. (1982). The Subject and Power. *Critical Inquiry* **8:** 777 – 95.

Gallagher, M. (2008). Foucault, Power and Participation. *International Journal of Children's Rights* **16:** 395 – 406.

Geberegziabher, S.F. (2006). Impact of Conditionality on Aid Effectiveness in Ethiopia. Paper delivered at International Economic History Congress, **31**.05.2006] Helsinki.

Gebreselassie, T. (2007). The Baseline Survey of the Occupational Safety and Health Conditions of Solid Waste Primary Collectors and Street Sweepers in Addis Ababa. Addis Ababa: International Labour Organization.

Gibbs, G. (ed.) (2007). *Analyzing Qualitative Data.* London: Sage.

Grindle, M. (ed.) (2007). *Going Local: Decentralization, Democratization, and the Promise of Good Governance.* Princeton: Princeton University Press.

Grindle, M. (2010). *Good Governance: The inflation of an idea.* Available at http://www.google.com/#hl=en&sclient=psy-ab&q=grindle%2C%20%20good%20governance%20the%20inflation%20&pbx=1&oq=&aq=&aqi=&aql=&gs_sm=&gs_upl=&bav=on.2,or.r_gc.r_pw.r_qf.,cf.osb&fp=e6372d9bb62944f8&biw=723&bih=751&pf=p&pdl=500. Accessed 25 February 2012.

Groves, L. (2004). Questioning, learning and "Cutting Edge" Agendas: Some Thoughts from Tanzania. In *Inclusive Aid. Changing Power and Relationships in International Development,* L. Groves and R. Hinton (eds). London: Earthscan.

Harold, M.G. (ed.) (1994). *A History of Ethiopia.* Berkeley: University of California Press.

Hinton, R. and L. Groves (2004). The Complexity of Inclusive Aid. In *Inclusive Aid. Changing Power and Relationships in International Development,* L. Groves and R. Hinton (eds). London: Earthscan.

Hoebink, P. (2006). European donors and "good governance": Condition or goal? *European Journal of Development Research* **18:** 131 – 61.

Hufty, M. (2009). *The Governance Analytical Framework.* Available at http://graduateinstitute.ch/webdav/site/developpement/groups/hufty_greg/public/Governance_Analytical_Framework.pdf. Accessed 13 February 2012.

Human Rights Watch (2010). Development without Freedom. How Aid Underwrites Repression in Ethiopia. New York : Human Rights Watch.

IGNIS (2009). *Income Generation and Climate Protection by Valorizing Municipal Solid Wastes in a Sustainable Way in Emerging Mega cities, Addis Ababa, Ethiopia*. Available at http://www.emerging-megacities.org/download/flyer-hauptphase/flyer_future_megacities_addis_ababa.pdf. Accessed 26 January 2012.

ILO (2007). Occupational Safety and Health of SWM Workers. Addis Ababa: International Labour Organization.

ILO (2008a). City wide Information, Education and Communication (IEC) Strategy for SWM. Addis Ababa: International Labour Organization.

ILO (2008b). *Employment Creation through Sustainable Solid Waste Management*. Addis Ababa: International Labour Organization. Available at http://www.ilo.org/public/english/region/afpro/addisababa/sro/pdf/solidwast.pdf.

ILO (2008c). Situation Analysis on Informal Economy Operators in Addis Ababa Waste Stream. Addis Ababa: International Labour Organization.

ILO (2008d). SWM Related Knowledge, Attitude and Practice (KAP) Analysis on Households. Addis Ababa: International Labour Organization.

ILO (2009). Situation Analysis on Informal Economy Operators in the Addis Ababa Solid Waste Stream. Addis Ababa: International Labour Organization.

Irish Aid (2008). Good governance, aid modalities and poverty reduction. From better theory to better practice. Final Synthesis Report for the Advisory Board of Irish Aid. London, ODI.

Jean-Pierre, and O.D. Sardan (2008). *Researching the practical norms of real governance in Africa* Available at http://www.institutions-africa.org/filestream/20090109-discussion-paper-5-researching-the-practical-norms-of-real-governance-in-africa-jean-pierre-olivier-de-sardan-jan-2009. Accessed 27 April 2012.

Kapuscinski, R. (2006). *The Emperor*. London: Penguin.

Karlson, S. (2012). *Good news from Africa: Finnish EKOKEM exports environmental expertise to Ethiopia*. Ministry for Foreign

Affairs of Finland. Available at http://formin.finland.fi/public/default.aspx?contentid=250175&nodeid=15148&contentlan=2andculture=en-US. Accessed 7 July 2012.

Lefort, R. (2010). *Ethiopia`s election: All losers*. Open Democracy. Available at http://www.opendemocracy.net/ren%C3%A9-lefort/ethiopias-election-all-losers. Accessed 0n 31 May 2012.

Meheret, A. (1999). The City of Addis Ababa: Policy Options for the Governance and Management of a City with Multiple Identities. Addis Ababa: Forum for Social Studies.

Myers, G.A. (2005). *Disposable Cities. Garbage, Governance and Sustainable Development in Urban Africa.* Aldershot: Ashgate.

Norconsult (1982). Addis Ababa solid waste management study. Phase 3, Master Plan. Addis Ababa.

ORAAMP (2002). Project proposal for Addis Ababa municipal solid waste management program. Addis Ababa: Office for Revision of Addis Ababa Master Plan.

Pausewang, S., K. Tronvoll and L. Aalen (2002). *Ethiopia since the Derg: A decade of democratic pretension and performance.* London: Zed Books.

Pierre, J. (ed.) (2000). *Debating Governance. Authority, Steering and Democracy,* Oxford: Oxford University Press.

Poluha, E. (2004). *The power of continuity: Ethiopia through the eyes of its children.*Uppsala: Nordic Africa Institute.

Postholm, M.B. (2005). *Kvalitativ metode. En innføring med fokus på fenomenologi, etnografi og kasusstudier.* Oslo: Universitetsforlaget.

PPIAF. (2011). *PPIAF Supports Private Sector Participation in the Solid Waste Sector in Ethiopia*. Available at http://www.ppiaf.org/sites/ppiaf.org/files/publication/PPIAF-Impact-Stories-PSP-in-Ethiopia-Solid-Waste-Sector.pdf. Accessed on 14 March 2013.

Santiso, C.(2001). Good Governance and Aid Effectiveness: The World Bank and Conditionality. Georgetown Public Policy Review, 7 (1):1-22.SBPDA (2003). Solid Waste Management

Status. Report of Addis Ababa: The way forward. Addis Ababa: Sanitation Beautification and Park Development Authority.

Schalkwijk, A. (2005). Is government open to participation of the poor? Urban policies and politics in Bolivia. In *Space and Place in Development Geography: Geographical Perspectives on Development in the 21st Century,* G. Nijenhuis, A. Broekhuis and G. Westen (eds). Netherlands: Dutch University Press.

Silverman, D. (2009). *A very short, fairly interesting and reasonably cheap book about Qualitative Research.* Thousand Oaks, CA: Sage.

Smith, B.C. (ed.) (2007). *Good Governance and Development.* Houndmills: Palgrave Macmillan.

Swyngedouw, E. (2000). Authoritarian governance, power, and the politics of rescaling. *Environment and Planning* **18:** 63 – 76.

Swyngedouw, E. (2005). Governance Innovation and the Citizen: The Janus Face of Governance-beyond-the-State. *Urban Studies* **42:** 191 – 206.

Søreide, T., A. Tostensen and I.A. Skage (2012). *Hunting for Per Diem. The Uses and Abuses of Travel Compensation in Three Developing Countries*. Oslo: Norad Evaluation Department.

Taye, A. and G. E.Tegegne(eds) (2007). *Decentralization in Ethiopia,* Addis Ababa: Forum for Social Studies.

Tronvoll, K. (2011). The Ethiopian 2010 Federal and regional elections: Re-establishing the one-party state. *African Affairs* **110** (438): 121-136.

UN Habitat (2002a). *The Global Campaign on Urban Governance*. Nairobi: United Nations Human Settlements Program.

UN Habitat (2002b). *Principles and realities of urban governance in Africa. UN Habitat-Global campaign on urban governance.* Nairobi: United Nations Human Settlements Programme

UN Habitat (2008). *Ethiopia Urban Profile*. Nairobi: United Nations Human Settlements Programme.

UN Habitat (2010a). *Collection of Municipal Solid Waste in Developing Countries*. Nairobi: United Nations Human Settlements Programme.

UN Habitat (2010b). *Solid Waste Management in The World's Cities. Water and Sanitation in the World`s Cities 2010*. London: United Nation Human Settlements Programme.

Vaughan, S. and K. Tronvoll (2002). *The Culture of Power in Contemporary Ethiopian Political Life*. Available at http://www.eldis.org/assets/Docs/14655.html. Accessed 31 January 2012.

World Bank (1992). *Governance and development.* Washington DC.

World Bank (2004). *Making services work for the poor*. New York and Oxford. Kpundeh, S and Khadiagala, G. M.(2008). *Information Access, Governance, and Service Delivery in Key Sectors: Themes and Lessons from Kenya and Ethiopia*. Premnotes, Public Sector Governance.World Bank: 121..

Yirgalem, M. (2008). The growth and management of Addis Ababa. Visible and invisible actors in urban management and emerging trends of informalization: A case study from Kolfe-Keranio subcity, Addis Ababa, Ethiopia. PhD thesis, Norwegian University of Science and Technology.

Zelalem, F.C. (2006). Controlling the Informal Sector: Solid Waste Collection and the Addis Ababa City Administration 2003-2005. Master's thesis, Norwegian University of Science and Technology.

Notes on Contributors

Dr Onyanta Adama holds a PhD in Human Geography. She is a senior researcher at the department of Human Geography, Stockholm University. At the time of writing the book she was a senior researcher at the Nordic Africa Institute, Uppsala, Sweden. Her primary research interest is urban governance and service delivery in African cities. Onyanta has conducted extensive research on solid waste management in Nigeria. She is particularly interested in how specific historical and political conditions combine with broader processes to shape governance structures and relations. The title of her PhD dissertation is *Governing from Above: Solid Waste Management in Nigeria's New Capital City of Abuja*. She has published in several international journals including Waste Management & Research, International Development Planning Review and Urban Forum. Adama can be reached at: onyanta.adama@humangeo.su.se

Mr. Olugbenga Adebola is the Managing Director and CEO of Richbol Environmental Services Limited, Lagos, Nigeria. He holds an MSc in Environmental Management and has over 20 years cognate experience in the field of waste and environmental management. He is involved in several Integrated Solid Waste and Environmental consultancies for various tiers of governments and parastatals, the private sectors, NGOs and Foundations in Nigeria. Adebola has several publications and research work to his credit, most of which were papers delivered at both local and international conferences and workshops. He is also a member professional bodies such as the Association of Waste Managers of Nigeria, International Collaborative Working Group (ICWG) on Solid Waste Management in low and middle income countries, and International Partnership for Expanding Waste Management Services of Local Authorities (IPLA), an initiative of United Nations Centre for Regional Development (UNCRD). Adebola can be reached at: adebolagb@richbol.com

Dr Adamu Ahmed is a Professor of Urban Planning in the Department of Urban and Regional Planning at the Ahmadu Bello University Zaria, Nigeria. He holds a first Degree in Geography and a PhD in Urban and Regional Planning. Ahmed is the Coordinator of the Center for Spatial Information Science and was formerly Head of the Department of Urban and Regional Planning at Ahmadu Bello University. His research interests include urban governance, urban growth dynamics and informality. Ahmed has also written extensively on growth management and participatory governance. His most recent publication is on ICT applications in monitoring urban growth and development control violations. Email: drahmedadamu@yahoo.com

Professor Raymond N.C Anyadike was a Professor in the Department of Geography, University of Nigeria, Nsukka. He supervised numerous Masters and PhD candidates in Climatology, Hydrology, Water Resources and Environmental Management many of whom are currently lecturers and researchers in Nigeria and overseas. Anyadike has to his credit over 60 publications in learned journals on aspects of climatology of West Africa, solid waste management and recycling, and water resources management. He was the author of the standard text *Statistical Methods for the Social and Environmental Sciences* and a co-editor of *Climate Change and the Nigerian Environment* both of which are widely used in Nigerian universities. Unfortunately, he died in August 2013 before the book project was completed.

Mr. Hammed Taiwo Babatunde has a BSc degree in Environmental Management and Toxicology from the University Agriculture, Abeokuta, Ogun State and a Masters in Public Health (Environmental Health) from the University of Ibadan, Nigeria. He is currently a PhD holder in the Dept. of Environmental Health Sciences, College of Medicine, University of Ibadan. He has worked in different capacities with NGOs that deal with environmental sanitation and management for over 7 years. Notable experience is being the Project Manager of a multi- million Naira project developed by (NINAAFEH/ MTN Nigeria Foundation Environmental Health Project) at Alesinloye market in Ibadan, Nigeria. Among other things, the integrated project comprises water supply and toilet facilities, solid waste recycling involving composting for organic wastes and conversion of plastic/nylon waste to pellets and resins and an incineration plant. He has also attended workshops at national and international levels and published in both local and international journals.

Dr Camilla Louise Bjerkli holds a PhD in Geography from the Norwegian University of Science and Technology (NTNU). The title of her PhD is: *Urban services and governance. The case of solid waste management in Addis Ababa, Ethiopia.* In her dissertation Bjerkli`s main interest were the implementation of various governance policies, in order to improve solid waste management and to explore how these policies were implemented in light of politics and power relations in the local context. Bjerkli`s main research interests includes urban informality and governance with focus on exploring and understanding the dynamics that occurs in the local context seen in light of culture, power and politics. Bjerkli works as a solid waste and environmental adviser at COWI AS. Bjerkli can be reached at: cbj@cowi.no

Mr. Emmanuel Onyekwere Chukwuocha is an Environmental Management and Development Consultant and founder EMOC Engineering and Environmental Services Limited Abuja, Nigeria. He holds an MSc degree in Environmental Management from Ondo State University, Akungba, Nigeria. He is the National Secretary of Waste Management Society of Nigeria and has over twenty-five years of varied experience. As former Head of Department of Waste Management and Sanitation in Abuja Environmental Protection Board, Chukwuocha initiated programmes and projects such as the public - private partnership in waste collection. He has served on several panels of Environmental Impact Assessment reviews of major national projects. Chukwuocha can be reached at: emmanuelchukwuocha@yahoo.co.uk

Dr. Chukwunonye Ezeah is a Chartered Waste Manager and a Fellow of the Higher Education Academy, UK. He is currently a Lecturer in the Faculty of Science and Engineering, University of Wolverhampton, United Kingdom. He graduated with a B.Eng (Hons) in civil engineering and holds a Master of Science Degree in Environmental Management and a Doctorate in Environmental Science from the University of Wolverhampton. His PhD thesis, *Analysis of Barriers and Success Factors Affecting the Adoption of Sustainable Municipal Solid Waste Management in Nigeria* helped inform recent policy decisions on waste management in Nigeria. Ezeah's research interests include sustainable waste management, environmental pollution and control, climate change, as well as environmental health and safety. His research publications have footprints in leading academic Journals such as *Waste Management, Journal of Solid Waste Technology and Management, Habitat International and Journal of Environmental Management* amongst others. Ezeah can be reached at: C.Ezeah2@wlv.ac.uk

Jak Fazakerley is an Environmental Consultant at Crestwood Environmental, UK. Fazakerley's work is focused within the waste and resource industry, particularly preparing and submitting Environmental Permit applications for organic waste, landfill and other waste treatment facilities, as well as monitoring landfill gas, leachate, groundwater, surface water, bio aerosols, odour and dust. Fazakerly graduated with a BSc degree in Geography and Environmental Management and an MSc in Environmental Technology, and is a graduate member of the Chartered Institution of Wastes Management (CIWM). Fazakerly's research interests include informal sector waste management, waste characterisation and compositional analysis, environmental pollution, monitoring and control. Email: jak@crestwoodenvironmental.co.uk

Idris Isah Iliyasu is a Lecturer in the Department of Urban and Regional Planning, Ahmadu Bello University, Zaria, Nigeria. He holds a Bachelor's Degree in Urban and Regional Planning and a Masters in Tourism and Recreation planning. His research interests are on urban quality and Livelihood and Governance. Iliyasu can be reached at: urbanist2005@gmail.com/iiiliyasu@abu.edu.ng

Professor Sridhar K. C. Mynepalli. Professor of Environmental Health Sciences at the Faculty of Public Health, College of Medicine, University of Ibadan, Ibadan, Nigeria. He obtained PhD from Indian Institute of Science, Bangalore, India, worked for 13 years years before moving to University of Ibadan, Ibadan in 1977. Worked on research projects ranging, from water resources, eutrophication and pollution control, water treatment, waste management, environmental toxicology, and community mobilization. Developed waste to wealth projects since 1980s particularly compost and organo-mineral fertilizers, biogas for communities using stakeholder approach. He established several pilot scale or demonstration units in communities. He was a consultant to UNDP, UNICEF, World Bank, several NGOs and Government bodies and agencies. Traveled extensively and visited several waste management facilities in India, Switzerland, Sweden, Finland, Denmark, and USA. Trained various categories in environmental field and published over 300 scientific papers in books and journals.

Dr. Chidozie Charles Nnaji is a Lecturer I at the Department of Civil Engineering, University of Nigeria, Nsukka. He holds a PhD in Water Resources and Environmental Engineering. His areas of research interest include: solid waste management, water quality monitoring, water treatment and distribution, wastewater treatment, hydrology, soil and groundwater pollution, environmental impact assessment, and rainwater harvesting and greywater recycling and re-use. Nnaji has been involved in some consultancy services in some of these areas and has authored about twenty articles in reputable peer reviewed journals. Nnaji can be reached at: chidozie.nnaji@unn.edu.ng

Dr. Thaddeus Chidi Nzeadibe is a Senior Lecturer and the Coordinator of Postgraduate Programmes in the Department of Geography, University of Nigeria, Nsukka. He earned a PhD in Environmental Management and has authored and co- authored articles on social participation in solid waste governance, urban poverty and livelihoods within the informal waste management system of Nigerian cities for *Applied Research in Quality of Life; Geography Compass; Habitat International; City, Culture & Society; Waste Management;* and *Environment, Development & Sustainability*. He was a Visiting Scholar at the Nordic Africa Institute (NAI), Uppsala, Sweden in 2013. His current research focuses on urban political ecology, exploring the interconnections of solid waste with urban governance, informal economy, and development in Africa. Nzeadibe can be reached at: chidi.nzeadibe@unn.edu.ng

Dr. Anthony C. Okoye is a Senior Lecturer in the Department of Environmental Management, Nnamdi Azikiwe University, Awka, Nigeria. He obtained a PhD in Environmental Chemistry and specializes in Pollution Monitoring and Waste Management. His areas of research interest include Climate Change, Environmental Impact Assessment and Energy. He is a certified waste manager. Okoye has more than forty five publications in both international and local journals and has attended several conferences in Nigeria and overseas. He is a Lead Fellow and a member of several societies. He is a registered Environmental Consultant. He was also the MD/CEO Anambra State Waste Management Authority, Nigeria. Okoye can be reached at: udicarl@yahoo.com

Dr Martin Oteng-Ababio is a Senior Lecturer and the Coordinator of the Urban Disaster Risk Reduction Programme in the Department of Geography and Resource Development at the University of Ghana. He holds a PhD in Geography from the University of Ghana and his dissertation was *Public-Private Partnership in Solid Waste Management in the Greater Accra Metropolitan Area*. He won the

Japanese Award for Outstanding Research on Development for 2011 and Best Paper Award at the Southern Africa Society for Disaster Reduction (SASDIR) Conference in 2012. Oteng-Ababio's research interests include urban environmental management; urban disaster risk reduction and informal urbanism. His research works have been published in reputable journals including *Habitat International, Environment and Planning A, International Development Planning Review, Applied Geography, Urban Geography*, and *Journal of Housing and the Built Environment*. Email: moababio@ug.edu.gh

Dr. Yusuf Rigasa Abdullahi is a Principal Lecturer at the Department of Environmental Science of Kaduna Polytechnic, Nigeria. He was a research fellow at University of Abertay Dundee, Scotland where he obtained his PhD in Waste and Environmental Management. Abdullahi is a researcher, a consultant and a social worker. His interest is in Environmental Performance and conservation of National Environmental Assets. Of special interest to him are the linkages and dynamics of communities and sustainable opportunities in the local environment. Abdullahi is on secondment to the National Oil Spill Detection and Response Agency (NOSDRA) as an Assistant Director in the Department of Oil Field Assessment (OFA). Abdullahi can be reached at: yusuf.rigasa@gmail.com

Dr Joy Ede Ukoje is a Lecturer I in the Department of Geography, Federal University Lokoja. She was the Coordinator of the Post Graduate Programme in the Department of Urban and Regional Planning at the Ahmadu Bello University, Zaria, Nigeria up to 2013. She holds a PhD in Urban and Regional Planning from Ahmadu Bello University, Zaria and her dissertation was *Analysis of Stakeholders Participation in Solid Waste Management in Zaria Urban Area*. Ukoje's research interests include urban environmental management; urban housing problems and informal sector activities. Her research works have been published both local and international journals. Ukoje can be reached at: joyukoje@yahoo.com

Index

Title Index

C

D

F

G

H

J

K

L

M

N

O

S

T

Y

Z

Author Index

A

B

C

D

E

F

G

H

N

O

W

Y

Z